LIBRARY/NEW ENGLAND INST. OF TECHNOLOGY
3 0147 1002 5332 0

D1524779

Design Guide for Composite Highway Bridges

Design Guide for Composite Highway Bridges

David C. Iles, The Steel Construction Institute

London and New York

First published 2001 by Spon Press
11 New Fetter Lane, London EC4P 4EE

Simultaneously published in the USA and Canada
by Spon Press
29 West 35th Street, New York, NY 10001

Spon Press is an imprint of the Taylor & Francis Group

Printed and bound in Great Britain by
St. Edmundsbury Press, Bury St Edmunds, Suffolk

British Library Cataloguing in Publication Data
A catalogue record for this book is available from the British Library

Library of Congress Cataloging in Publication Data
A catalog record for this book has been requested

ISBN 0-415-27453-2

CONTENTS

ACKNOWLEDGEMENTS

Design Guide for Composite Highway Bridges is an amalgamation of guidance and example calculations previously given in two separate Steel Construction Institute publications. Both publications prepared by Mr D.C. Iles of the Steel Construction Institute, with advice from members of the Steel Bridge Group.

The Steel Construction Institute expresses thanks to Messrs Bullen and Partners, the Travers Morgan Group and Cass Hayward and Partners, for permission to use the designs shown in the worked examples. Thanks are also expressed to Mr C.R. Hendry (W.S. Atkins) and Miss N. Knudsen (SCI) for carrying out checks on the worked examples.

The work leading to this publication was funded by Corus (formerly British Steel).

1 INTRODUCTION

Composite construction, using a reinforced concrete slab on top of steel girders, is an economical and popular form of construction for highway bridges. It can be used over a wide range of span sizes.

Design Guide for Composite Highway Bridges covers the design of continuous composite bridges, with both compact and non-compact sections, and simply supported composite bridges of the 'slab-on-beam' form of construction.

The guide assumes that the reader is familiar with the general principles of limit state design and has some knowledge of structural steelwork. It provides advice on the general considerations for design, advice on the initial design process and detailed advice on the verification of structural adequacy in accordance with BS 5400. It concludes some advice on structural detailing. The determination of design forces throughout the slab is described, key features relating to slab design are identified, and detailed advice on slab design is given. The selection of protective treatment and bearings is excluded, being well covered in other texts[1][2].

This guide includes a set of twelve flow charts that summarise the design process following the rules in BS 5400, taking into account the significant amendments recently made in the latest issue of BS 5400-3 *Code of Practice for the design of steel bridges*[3].

Three worked examples describe the initial and detailed design aspects for a four-span bridge, a three-span bridge and for the deck slab of a simply supported bridge. Each example is presented as a series of calculation sheets, with accompanying commentary and advice given on facing pages.

Where reference is made to a clause in one of the Parts of BS 5400, the reference is given in the form '3/9.1.2', which means clause 9.1.2 of BS 5400-3.

References are made in the text to further advice in 'Guidance Notes'. These are a series of notes, published by The Steel Construction Institute[4], that give concise advice from the members of the Steel Bridge Group, a technical group of experienced designers, fabricators and clients. A full list of Guidance Notes is given in Appendix A.

2 DESIGN CODES

2.1 BS 5400

The design and construction of continuous composite bridges is covered by British Standard BS 5400: *Steel, concrete and composite bridges*[5]. The document combines codes of practice to cover the design and construction of bridges and specifications for the loads, materials and workmanship. It is based on the principles of limit state design. It comprises the following Parts:

Part 1	General statement
Part 2	Specification for loads
Part 3	Code of practice for design of steel bridges
Part 4	Code of practice for design of concrete bridges
Part 5	Code of practice for design of composite bridges
Part 6	Specification for materials and workmanship, steel
Part 7	Specification for materials and workmanship, concrete, reinforcement and prestressing tendons
Part 8	Recommendations for materials and workmanship, concrete, reinforcement and prestressing tendons
Part 9	Bridge bearings
Part 10	Code of practice for fatigue

The general principles of the limit state design approach are given in Part 1. Part 1 states that two limit states are adopted in BS 5400, the ultimate limit state (ULS) and the serviceability limit state (SLS). The criterion for structural adequacy is expressed as:

$$R^* \geq S^*$$

i.e. the design resistance R^* (based on nominal strength divided by partial factors) shall be at least equal to the design load effects S^* (based on nominal loads multiplied by other partial factors) - see Clauses 1/2.3 and 1/5.

Part 2 specifies loads that are to be taken into account in the design. Parts 3, 4, 5 and 10 are Codes, which are manuals of good practice for the design of bridges. Implicit in the Codes is the assumption that workmanship and materials will be in accordance with the Specifications of Parts 6 and 7. These two Parts are written in a form suitable for incorporation in contract documents. In particular, Part 6 provides a comprehensive specification for the various forms of steel (plates, sections, bolts, welds, etc.) and the quality of workmanship employed in fabrication and erection.

In Part 3 (as amended in 2000), reference is made to a number of product standards for the steel material, the most commonly recognised of which is BS EN 10025. Part 6 makes reference to these and other supporting standards for materials, workmanship, inspection and testing, etc. See further comment in Section 2.6.

2.2 Interrelation of Parts 3, 4 and 5 of BS 5400

Part 5 of BS 5400 deals with the design of composite bridges, but provides detailed requirements only for the interaction between steel and composite elements. Design of the separate elements is referred to Parts 3 and 4 of the Standard.

A particular point to note when using Parts 3, 4 and 5 is the different way in which the partial factor γ_{f3} is applied. In Part 3 (steel bridges) the calculated strengths are divided by this factor for comparison with the load effects, whereas in Part 4 (concrete bridges) the strengths are compared against the load effects multiplied by the factor. Care must therefore be taken when applying γ_{f3}. It is suggested that γ_{f3} is consistently applied on the strength side for all elements of composite structures, both steel and concrete; this approach has been adopted for the Worked examples [1].

Attention is also drawn to the different treatments of the partial factor γ_m. Part 3 gives values for γ_m that are to be applied in various circumstances to expressions for design strengths (resistances); the factor is explicitly included. In Part 4, γ_m is often implicitly included in expressions for design strength (such as ULS moment resistance of a slab).

2.3 Design Manual for Roads and Bridges

The Design Manual for Roads and Bridges (DMRB) comprises a collection of Standards and Advice Notes issued by the 'Overseeing Organisations' [The Highways Agency (for England), The Scottish Executive Development Department (for Scotland), The National Assembly for Wales (for Wales) and The Department for Regional Development (for Northern Ireland)]. The collection includes some documents issued before 1992 by the Department of Transport that are still valid.

In relation to bridges, the documents give guidance to the designer and provide interpretation and application of BS 5400. They also correct typographical errors in the Standard and amend it where considered appropriate. A list of the key documents relating to the design of new bridges is given in Appendix B. Designers should check that they have up-to-date copies when carrying out design.

2.4 Specification for Highway Works

The four Overseeing Organisations also issue the *Manual of Contract Documents for Highway Works* (MCDHW), which comprises six separate Volumes. These documents provide the basis for documentation for individual contracts, and are supplemented, for each contract, by project-specific requirements. Of particular relevance to steel bridge construction are the sections known as 'Series 1800' of Volume 1, *Specification for Highway Works* (SHW), and Volume 2, *Notes for Guidance on the Specification for Highway Works*. The SHW implements BS 5400-6, modifies some of its clauses and provides the framework for additional project-specific requirements. For guidance on the latter, see SCI's *Model Appendix 18/1* document [6].

2.5 Requirements of Railtrack

Highway bridges that cross over a railway line need to meet the requirements of Railtrack.

Railtrack issues Railway Group Standards that set out its requirements for the design of structures. It also issues a Code of Practice document. Generally, the requirements follow those in the DMRB, but with a few detailed variations and with more extensive requirements for railway loading and for accidental loading (on both underline and overline bridges). The relevant documents are listed in Appendix B.

Railtrack also issues model specification clauses that amplify the specifications in BS 5400, in a similar way to the SHW.

2.6 European Standards

In recent years, many British Standards (issued by the British Standards Institution) have been replaced by European Standards. These European Standards have been developed by subcommittees of the European standards body, known as CEN. BSI is a member of CEN and now issues the CEN standards in the UK under the designation BS EN (followed by the reference number). Perhaps the most obvious BS EN documents for a steel bridge designer are those relating to the steel material itself; BS 4360[7], which was referred to by BS 5400-3:1982, has been replaced by a series of BS EN documents[8][9][10]. Other standards, including those relating to matters such as welding, are in the process of being replaced by CEN standards.

2.7 References to BS 5400 in this publication

References are made in this publication to clauses in various Parts of BS 5400. For simplicity, these are expressed in the fashion 3/4.5, where the 3 refers to Part 3 and the 4.5 refers to the clause number in that Part. A list of all the Parts and the dates of the versions current at the time of publication are given in Appendix B.

However, two special cases must be noted. Part 2 (Specification for loading) has been revised by BD 37/88 (see the DMRB) and a completely new version is included in an Appendix to that document. Because the BD applies to most bridges in the UK, all subsequent references to Part 2 in this publication are to the BD 37/88 version, unless noted otherwise.

Similarly, BD 16/82 made significant changes to Part 5 (Design of composite bridges) and a separate 'combined' document (Part 5 plus the BD revisions) was produced by the (then) Department of Transport. (The latter document is sometimes known as the 'yellow document', because it was issued with a yellow cover.) Again, all subsequent references to Part 5 in the present document are to the Part 5 as amended by BD 16/82, unless noted otherwise.

3 CHARACTERISTIC FEATURES OF COMPOSITE BRIDGES

3.1 Beam and slab construction

The form of construction considered in this publication is the beam and slab type, where a reinforced concrete deck slab sits on top of several I-section steel girders, side-by-side, and acts compositely with them in bending. It is one of the most common types of recent highway bridge in construction in the UK. A typical cross section, for a two-lane road with footways, is shown in Figure 3.1.

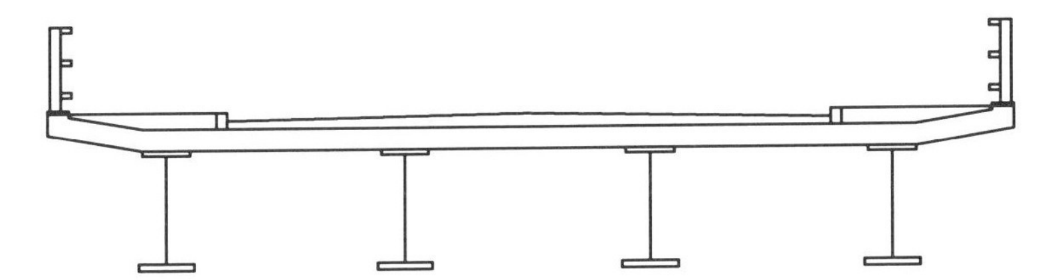

Figure 3.1 *Typical cross section of a composite highway bridge*

Composite action is generated by shear connectors welded on the top flanges of the steel girders. The concrete slab is cast around the connectors. This effectively creates a series of parallel T-beams, side by side. The traffic runs on a non-structural wearing course on top of the slab (there is a waterproofing membrane between). The load of the traffic is distributed by bending action of the reinforced concrete deck slab, either transversely to the longitudinal beams or, in some cases, by longitudinal bending to cross-beams and thence transversely to a pair of longitudinal main beams.

The steel girders can be of rolled section, for fairly short spans, or can be fabricated from plate.

Rolled sections – Universal Beams

Universal Beams are available up to 1016 mm deep (sizes above 914 mm are outside the range in BS 4[11]) and may be used for spans up to about 25 m for a simple span and up to about 30 m for continuous spans. Greater spans can be achieved if the bridge is lightly loaded – a farm access bridge or a footbridge, for example. In both the latter cases, where the beam is shallow relative to the span, considerations of deflection and/or oscillations may control the design.

Very little fabrication is necessary with Universal Beams, usually only the fitting of stiffeners over support bearings and the attachment of bracing. Beams can be curved in elevation (camber) by specialist companies using heavy rolling equipment.

Plate girders

For highway bridges where spans exceed the limits dictated by the maximum size of Universal Beams, girders must be fabricated from plates. Even for smaller spans, plate girders may be more suitable, because thicker webs and flanges can be provided. Also, Universal Beams of 762 mm serial size and above can often be more economically replaced by a similar plate girder.

The use of plate girders gives scope to vary the girder sections to suit the loads carried at different positions along the bridge. A wide variety of different forms in elevation and section has developed.

The designer is free to chose the thickness of web and size of flange to suit the design, though it must be remembered that too many changes may not lead to economy, because of the additional fabrication work. Splices are expensive, whether bolted or welded. Except for long span structures, the most economical solution is usually to put splices where changes of plate thickness occur at or near the point of contraflexure.

With plate girders, the designer can also choose to vary the depth of the girder along its length. It is quite common to increase the girder depth over intermediate supports. For spans below about 50 m, the choice (constant or varying depth) is often governed by aesthetics. Above 50 m, varied depth may offer economy because of the weight savings possible in midspan regions. The variation in depth can be achieved either by straight haunching (tapered girders) or by curving the bottom flange upwards. The shaped web, either for a variable depth girder or for a constant depth girder with a vertical camber, is easily achieved by profile cutting during fabrication.

Beam/girder spacing

Typically, girders are spaced between about 3.0 and 3.5 m, and thus, for an ordinary two-lane overbridge, four girders are provided. This suits the deck slab (typically 240 mm thick), which has to distribute the direct loads from the wheels.

When the spans exceed about 50 m, an alternative arrangement with only two main girders is sometimes used. Then the slab is supported on cross-beams, which now span transversely between the two main girders; the slab spans longitudinally between cross-beams, rather than transversely between main girders. This is sometimes referred to as 'ladder beam' construction, because of the plan configuration of the steelwork. (See further comments in Section 4.)

Beams need to be braced together at support positions, for stability and to effect the transfer of horizontal loads (wind and skidding forces) to the bearings beneath the girders. Bracing is often needed at discrete positions in the spans, to stabilise compression flanges; plan bracing may also be needed for the construction condition.

Box girder bridges

Composite bridges can also be constructed using steel boxes with a reinforced concrete deck. The steelwork may be in the form of either closed boxes or open topped boxes (which are closed by the addition of the slab). Box girder bridges are outside the scope of this publication; guidance is available in a separate SCI publication[12].

Welding

Welding is a necessary part of the construction of every modern steel bridge. Welding technology is a specialist field, but designers need to understand how to design welded connections and the practical considerations in carrying out the welding.

3.2 Forms of slab construction

The most usual form of slab construction is a uniform thickness of slab cast in situ. With fairly narrow bridges (e.g. two lanes plus footways), all the beams can be at the same level and the road camber achieved by regulating the thickness of surfacing (see Figure 3.1). With wider bridges and bridges where there is a significant crossfall (such as superelevation on a curve), the beams will be at different levels; the slab top surface will follow the required road profile, the soffit may either be parallel to it (which requires a small haunch or step on the uphill sides of the top flange) or may be aligned with the edges of the top flanges (the slab will thus be slightly thicker on the uphill edge of the flange). Haunched slabs and tapered thickness slabs are sometimes used in conjunction with wider-spaced main beams or central stringer beams (see Section 4.3).

Temporary soffit formwork, usually timber, can be used in conjunction with all types of slab for in-situ construction.

Permanent formwork can be used with *in-situ* slab construction to speed construction and reduce costs where the span of the slab does not exceed about 4 m and the soffit is plane. One common type of permanent formwork is a proprietary precast plank of high grade concrete, suitable for spanning between main girders and acting compositely as part of the deck slab. Projecting reinforcement on the upper face keys it to the in-situ slab, so that the two act integrally, as well as providing truss-like action to carry the wet concrete. Glass reinforced plastic (GRP) permanent formwork with steel ribs (enclosed within the GRP) is another attractive option, though this does not participate structurally with the slab. Durability is a prime concern with permanent formwork, both in terms of fatigue and corrosion. The Highways Agency has carried out tests on various types of permanent formwork, and their comments are given in Advice Note BA 36/90 (see Appendix B). The use of permanent formwork on UK trunk roads currently requires acceptance by the relevant Technical Approval Authority.

Instead of in-situ construction, full thickness sections of precast slab can be used in certain circumstances. This may require suitable provision of pockets for shear connectors and would thus affect detailing considerably. It is particularly worth mentioning that the use of precast cantilever slabs can reduce the problem of formwork support considerably.

3.3 Shear connection

There are a number of ways of providing a shear connection between the steelwork and the deck slab but by far the most common is the stud connector. This is essentially a headed dowel that is welded to the top flange using a special tool that supplies an electrical pulse sufficient to fuse the end of the dowel to the flange.

Alternative forms of shear connectors include steel bars with hoops, and short lengths of channel. Although both of these provide a higher resistance per unit, they each have to be welded on manually and consequently are more expensive.

3.4 Substructures, expansion joints and integral bridges

Most composite bridges have traditionally been restrained horizontally in a 'determinate' manner on bearings on top of abutments and intermediate supports. The superstructure is then free to expand and contract, and horizontal forces are resisted in such a way that the forces on the individual bearings can be calculated by statics. The arrangement of fixed, free and guided bearings is known as the 'articulation', and further advice on this aspect is given in Guidance Note 1.04. The design of substructures is outside the scope of this publication.

The freedom of the deck to expand and contract requires expansion joints at roadway level. Experience over many years has shown that these joints are a source of maintenance problems, in that they require attention and replacement, because of wear and, even more significantly, they can allow dirt and salty run-off from the roadway to penetrate to the structure and to the bearings, causing extensive deterioration. This experience has led to the consideration of 'integral bridges', where there is no joint at roadway level and the support structure is forced to displace with the movements of the deck. The subject of integral bridges is a separate topic that is covered by a series of SCI publications[13][14][15], although the guidance in this publication is still relevant for the detailed design of the superstructure.

3.5 Corrosion protection

The steel girders are usually given a high standard of protective treatment against corrosion. This treatment consists of abrasive blast cleaning followed by the application of multiple coats of paint or similar material (aluminium spray is often specified as the first coat in the coating system). Requirements for coating systems are given in the 1900 series of the *Specification for Highway Works* and guidance can be found in *Steel highway bridges protection guide*[1].

As an alternative, in suitable environments, a special weather resistant steel may be used. This requires no protective treatment, because it forms a stable layer of oxidised material (known as a 'rust patina'). See advice in Guidance Note 1.07.

4 INITIAL DESIGN

4.1 General

In this guide, it is assumed that the initial design is carried out when the road layout, disposition of traffic lanes, footways, parapets, etc. have already been determined. Spans may have been determined by topographical or other constraints largely outside the control of the designer.

The initial design stage is considered here to cover the preliminary selection of structural arrangement and member sizes, without the need for extensive calculation. It is followed by the detailed design stage (Sections 5, 6 and 7 of this guide), which covers the detailed calculation in accordance with the Code, and leads to confirmed structural arrangements and details, ready for the production of construction documents.

A value for the construction depth may already have been assumed in determining road levels and thus in assessment of earthworks quantities. This should be reviewed in line with appropriate span/depth proportions for composite construction (see Section 4.4).

Selection of an appropriate girder spacing will be made during the initial design (see Section 4.3), unless it has already been determined by other constraints, such as positioning of columns or form of pier at intermediate supports.

4.1.1 Simple spans

Simple spans are used where it is unnecessary or uneconomic to use more than one span, or where it is desirable to eliminate midspan piers, or where significant settlement of supports can be expected.

Short simple spans are often associated with limited clearances, such as over a minor river; although concrete bridges often fill this role, a composite beam and slab bridge can offer a competitive alternative.

Medium length spans of around 30 m can be used to clear a dual carriageway without intermediate support, and if designed to shallow depth/span proportions can provide a clean, light appearance. Over a dual four-lane motorway, a span of just over 40 m would be required.

Long simple spans, over 50 m, are used only occasionally; such crossings usually involve adjacent spans and continuous construction.

4.1.2 Continuous spans

Where multiple spans are required, continuous construction is preferable for economy and durability (i.e. to avoid joints at intermediate supports and the associated difficulties of dealing with roadway run-off).

Where the designer is free to select or vary the span lengths in the initial design, the following points should be considered:

- the sizes that can be conveniently fabricated and erected
- economy in repetition (where there are many spans)
- the size of end spans (ideally 80–85% of internal spans)
- favourable positioning of site splices.

The selection of span lengths may be influenced by an economic balance between substructure and superstructure costs. As spans increase from about 35 m, the cost of steel superstructure steadily increases in relation to the cost of the substructure. It must be pointed out, however, that the balance between the two cost elements is not the same for steel as it is for concrete construction: *economic spans to suit concrete are not necessarily the same as those to suit composite construction.* Span lengths should be reviewed carefully if a composite design is being considered as an alternative to an all-concrete design.

4.1.3 Skew bridges

Constraints on roadway alignment are increasingly leading to the requirement for bridges that are significantly skew to the features that they cross. The support of beam and slab bridges on lines that are highly skewed (i.e. a large angle between the line of support bearings and a line square to the main beams) leads to complications, particularly in the detailing of bracing. The consequences of a skew alignment should be considered at an early stage of the initial design. For further advice, see Guidance Note 1.02.

4.1.4 Integral bridges

The DMRB requires, in BD 57/95, that integral construction be considered for all bridges up to 60 m overall length and where the skew does not exceed 30°. This length is not a limit and integral construction can be used for longer bridges. For guidance on the use of integral construction, see the series of SCI publications on this subject [13].

4.2 Loadings

Highway bridges are usually designed to carry a combination of uniformly distributed loading representing normal traffic (type HA) and an abnormal heavy vehicle (type HB). These loads, together with other secondary loads, are specified in Part 2 of BS 5400, except that the magnitude of the abnormal vehicle is chosen to suit the particular requirements for the road (usually 30, 37.5 or 45 units of loading).

As mentioned in Section 2.7, it should be noted that BD 37/88 has modified the loading specification (particularly the HA loading), and the version of Part 2 in the Appendix to BD 37/88 should normally be used.

In addition, the highway authorities require the consideration of Abnormal Indivisible Loads (AIL) on routes designated as Heavy or High Load Routes. In some cases, these loads will have a more severe effect on the superstructure than HB loading.

Where the clearance over a highway is less than 5.7 m, the DMRB requires that the superstructure be designed for collision loads, in accordance with BD 60/94. These loads are onerous, particularly for small beams. It is better to provide sufficient clearance and thus avoid the loading.

4.3 Girder spacing and deck slab thickness

The form of construction usually employed for simple spans, and for spans up to about 45 m in continuous bridges, consists of a number of I-girders acting compositely with a reinforced concrete deck slab supported on the top flanges of the steel girders.

In addition to its role in augmenting the bending resistance of the girders, the deck slab distributes wheel loads to the main girders and transfers some load from more highly loaded girders to adjacent girders. The spacing of main girders thus affects the design of the slab, as well as determining the number of girders required.

Wheel loads usually determine the design of the slab, including its reinforcement; the minimum thickness of slab (required to carry global dead and live load effects, plus either local HA or local HB loads) is about 220 mm, on the basis of current cover and crack width requirements.

The total transverse moments in the slab are not particularly sensitive to girder spacing in the range 2.5 to 3.8 m (the increase in local moment with span is more or less balanced by a reduction in the moment arising from the transfer of load from one girder to the next). The optimum slab thickness overall is typically 230-250 mm. It is advantageous to choose a spacing as high in the range as possible, consistent with other geometrical considerations.

In selecting a suitable girder spacing, attention must be paid to the cantilevers at the edges of the deck. These should normally be restricted to about 1.5 m. Whilst the cantilever could be increased to about 2.5 m if it carries a footway that is protected by a crash barrier (thus avoiding local accidental vehicle loading), support of falsework during construction also tends to lead to a practical limit of about 1.5 m.

Where high containment barriers are specified, the length of cantilevers may need to be restricted, and the slab increased in thickness locally.

For reasons discussed in Section 5.3.7, it is often preferable, where possible, to use an even number of girders so that they may be paired during construction.

For continuous bridges with spans from about 45 m to over 100 m, alternative arrangements with just two principal plate girders are often used to avoid the rather uneconomic use of multiple webs. Such arrangements include:

(i) Haunched slab

Haunching of the slab close to the plate girders (to about 350 mm thickness) can be used to increase girder spacing up to about 6 m. In Europe, haunching is used in conjunction with transverse prestressing, but this is rarely employed in the UK.

(ii) Intermediate stringer beam

A single Universal Beam midway between main girders, supported by cross bracing at regular intervals, can be used to increase main girder spacing to about 7 m. This arrangement has been used on several bridges in the UK. It requires extra attention to bracing design, detailing and fatigue resistance.

(iii) Cross girders

Widely spaced main girders (over 7 m) can be used in conjunction with cross girders spaced 3.5 to 4.0 m apart. The slab spans longitudinally between cross girders, and the cross girders, acting compositely with the slab, span transversely between main girders. This form is sometimes known as ladder deck construction, because of the arrangement of steelwork in plan. The changes to HA loading over short lengths that have been made by BD 37/88 add significantly to the design load of the cross girders. The ladder deck form is best suited to bridge decks about 12 m in width.

These three forms are shown in Figure 4.1.

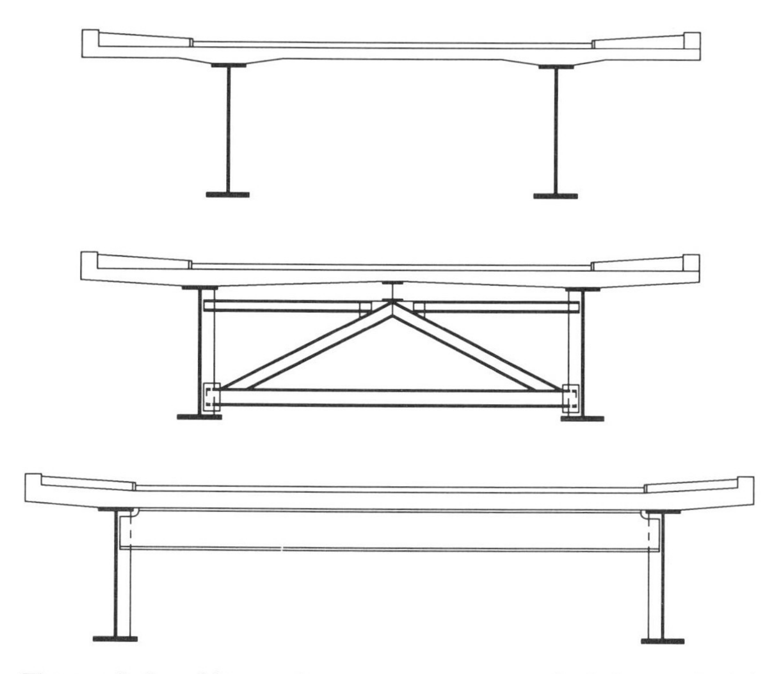

Figure 4.1 *Alternative arrangements of girder and slabs*

It should be noted that the cost of the slab may be over 25% of the total superstructure cost, and the investigation of more than one slab thickness and more than one arrangement of girders is worthwhile for larger spans and for multiple spans.

4.4 Construction depth

For simple spans over about 25 m, a construction depth (top of slab to underside of beam) of between about 1/18 and 1/30 of the span can be achieved with fabricated beams, though the most economic solution will be toward the deeper end of this range. For shorter spans the depth is likely to be proportionately greater, particularly when using Universal Beams for spans under 20 m.

For composite continuous spans with parallel flanges, the construction depth (again, from top of slab to underside of beam) is typically between 1/20 and 1/25 of the major span. The use of curved soffits or tapered haunches can reduce construction depth at midspan, at the expense of increased depth at the internal supports. A selection of typical arrangements is given in Figure 4.2.

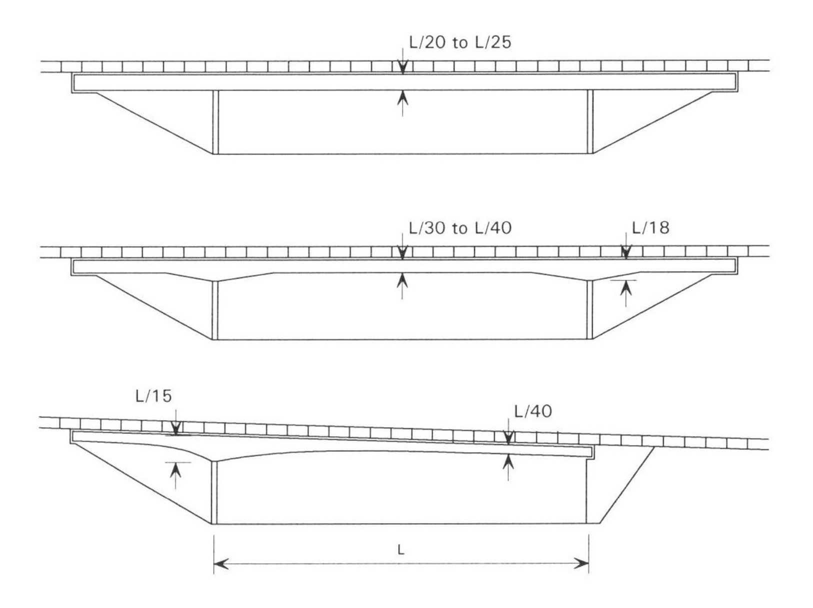

Figure 4.2 *Typical span/depth proportions for continuous spans (depths from top of structural slab)*

4.5 Initial selection of flange and web sizes

Experience and a few personal rules of thumb can often be used for an initial selection of sizes. Because the weight of the steelwork contributes little to overall design moments, the selection can be quickly refined. One such set of simple rules is given in Appendix C.

The British Steel (now Corus) publication *Composite steel highway bridges* [16] gives some guidance on the choice of cross section for fabricated continuous beams. It is intended for use with non-compact sections of between 20 and 60 m span. It also gives guidance for simple spans using Universal Beams, for both compact and non-compact section design.

Spans that must be fabricated in several lengths give the opportunity to vary the girder make-up in the different pieces required for each span. Maximum length of the pieces is usually governed by either transportation (loads over 27.4 m long require special arrangements) or by length of plate available from the mills (see Section 4.7 below).

The main structural steel members in bridges are usually grade S355, which is more cost effective than grade S275. Higher strength steel grades (S420 and S460) are available and can be used in designs to BS 5400 but they are more expensive and have not yet been used in the UK to any significant extent in bridgework.

Selection of the particular subgrade (for toughness or other requirements) does not usually need to be considered in the initial design unless availability is likely to be a problem. Advice about the steel materials available and the choice of subgrade, etc. is given in Guidance Note 3.01.

Universal Beams are a natural choice for the girders of shorter spans; Universal Beams should be selected in the initial design for simple spans up to about 20 m (18 m if 45 units of HB loading are to be carried) and for continuous spans up to about 25 m. However, the economies offered by fabricated sections of similar proportions should then be considered carefully in the detailed design stage.

As mentioned in Section 3.5, weather resistant steel can be used in suitable environments. Use of weather resistant steel requires that a 'corrosion allowance' is made on all exposed surfaces; this allowance should be made at the initial design stage, as well as in the detailed design. Corrosion allowances are specified in Clause 3/4.5.6 and BD 7/81 (see Appendix B).

4.6 Choice of compact or non-compact sections

Nearly all structural configurations of composite bridges that are likely to be considered for simple spans and for midspan regions of continuous spans will be of compact proportions, because the slab in compression keeps the plastic neutral axis high (often in the flange, or just in the web).

For continuous construction, the intermediate support regions of composite beams will usually be non-compact, again because the plastic neutral axis is high, which now means that a significant part of the web is in compression.

In the initial selection of girder size, it would be reasonable to assume that midspan regions will be designed on the basis of compact sections, but it may be preferable to base the selection simply on the non-compact moment resistance, which leaves some margin for additional capacity (or for a small reduction in beam size) in the detailed design.

4.7 Availability of steel beams and plates

Steel material, in the form of plates and sections, is manufactured in accordance with various European Standards, the most relevant to bridgework being:

BS EN 10025 Hot rolled products of non-alloy structural steels.

BS EN 10113 Hot rolled products in weldable fine grain structural steels.

BS EN 10155 Structural steels with improved atmospheric corrosion resistance.

As-rolled plate sizes (length × width × thickness) are determined by the size and weight of slab used when rolling from slab to plate. In general, the maximum plate length produced by Corus is 18.3 m, but longer lengths are available by agreement. Strength grades up to S460, to the above Standards, are available.

Corus rolls Universal Beams up to a maximum size of 1016 × 305 × 487 kg/m. Strength grades up to S355, to the above Standards, are available.

Full details of the sizes available from Corus are given in separate publications[17][18]. The designer is advised to consult Corus on the availability of maximum plate sizes if the design requires their use.

For further advice on selection of steel material, see Guidance Note 3.08.

4.8 Economic and practical considerations

4.8.1 General considerations

Clean lines to the overall appearance and minimum use of complex details are most likely to lead to an economic and efficient bridge structure, though external constraints often compromise selection of the best structural solution.

The fabrication of the basic I-section is not particularly expensive, especially with the use of modern semi-automatic girder welding machines (T and I machines). Overall fabrication cost is of the same order of cost as the material used. With the widespread use of computers in design and in control of fabrication shop machines, geometrical variations, such as curved soffits, varying superelevation and precambering, can be readily achieved with almost no cost penalty. Much of the total cost of fabrication is incurred in the addition of stiffeners, the fabrication of bracing members, butt welding, the attachment of ancillary items, and other local detailing that leads to a significant manual input to the process. The designer can thus exercise freedom in the choice of overall arrangement but should try to minimise the number of small pieces that must be dealt with during the fabrication process.

Expert advice should be obtained from fabricators to assist in the choice of details at an early stage in the design. Most fabricators welcome approaches from designers and respond helpfully to any interest shown in their fabrication methods.

The form of the substructure at intermediate supports, whether for reasons of appearance or of construction, often has a strong influence on the form of the superstructure. For example, a low clearance bridge over poor ground might use multiple main girders on a single broad pier, whereas a high level bridge of the same deck width and span over good ground might use twin main girders, with cross girders, on individual columns.

Highly skewed bridges are sometimes unavoidable but it should be noted that the high skew leads to the need for a greater design effort, more difficult fabrication and more complex erection procedures. In particular, the analytical model, the detailing of abutment trimmer beams, precambering and relative deflection between main beams must all be considered carefully.

4.8.2 Construction considerations

Construction of a composite bridge superstructure usually proceeds by the sequential erection of the pieces of the main girders, usually working from one end to the other, followed by concreting of the deck slab and removal of falsework. However, situations vary considerably and constraints on access are a major influence on the erection sequence for any bridge. In some cases they might determine the form of the bridge. It has always been good practice in most cases to ensure that, before proceeding to detailed design, at least one erection scheme is examined and the requirements for it included in the detailed design.

This good practice is now emphasised by the Construction (Design and Management) Regulations, which require a designer to anticipate how a structure will be built and to assess the risks involved in the construction. See further comment in Guidance Note 9.01.

In some circumstances, where access from below is difficult or impossible, launching from one or both ends may be appropriate. If so, this is likely to have a significant effect on girder arrangements and detailing. Advice should be sought from an experienced contractor.

Stability of girders during erection and under the weight of wet concrete will have a significant effect on the size and bracing of the top flange in midspan regions and, to a lesser extent, on the bracing to the bottom flange adjacent to intermediate supports. Temporary bowstring bracing to individual girders may need to be provided if they are too heavy to be erected as braced pairs.

Site splices between the main girder sections are commonly connected with high strength friction grip (HSFG) bolts. Welded joints are more expensive and prove more onerous on quality control on a small job but should be considered on larger jobs and where their better appearance is warranted. One method or the other should be adopted throughout the bridge; it is uneconomic to use both methods.

Transportation by road imposes certain limitations on size and weight of fabricated assemblies. The most frequently noted limitation is a maximum length of 27.4 m, above which special notification and procedures apply. Nevertheless, UK fabricators are used to transporting longer loads - in exceptional cases girders well over 40 m long have been transported. See further comment in Guidance Note 7.06, and seek the advice of a fabricator when considering the use of large or heavy components.

4.8.3 Maintenance considerations

Bridges are expected to have a long life (the design life given in BS 5400 is 120 years) but will certainly require some maintenance during their lives. The design, and in particular the detailing, should recognise the need for durability and facilitate whatever maintenance will be necessary. To maximise the life of protective coatings, avoid any features that would trap water or dirt, and ensure adequate access for the proper application and maintenance of the coatings. Access for maintenance should either be provided or be possible with the minimum of temporary works. See further advice in SCI publication P154[19].

It is likely that bearings, particularly sliding bearings and elastomeric bearings, will need to be replaced during the life of a bridge. Provisions need to be made in the design for temporary support during replacement.

The CDM Regulations also require the assessment of hazards during maintenance work. The design should avoid or reduce, as far as practicable, risks during maintenance.

4.8.4 Future developments

Current and future developments in steel production and fabrication may well affect the form of design for composite bridges. Fabricators are now using semi-automatic girder welding machines (T and I machines) and automatic girder drilling machines. These are best suited to parallel flange, constant overall depth, constant width flanges; designs might in future be tailored to make the best use of these machines.

Whilst the cost of manually fitting and welding web stiffeners in the shops has favoured the use of slightly thicker webs with fewer stiffeners, the increasing use of robotics in the fabrication shop, particularly in positioning and welding web stiffeners, might affect the choice of stiffener types, arrangements and details in future.

Greater speed of construction is achieved with precast slabs. Currently by far the greatest part of the site operations on the superstructure involves in-situ slab construction. Pressure for shorter construction times (and lower costs) may lead to further developments in the use of precast slabs. Several forms of full-thickness precast slab units have been used recently in Sweden; a bridge in Ireland has recently been built to BS 5400 with full thickness deck units on top of a four-girder I-beam configuration.

Lightweight concrete offers the benefit of lower dead load (slab density 70% that of normal concrete, or less), which may be advantageous for longer spans. It is used in Europe but has not yet been used much in the UK for bridge decks, although there have been some cases where it has been used to advantage. This seems to be because of reservations about workability and quality control during casting. Developments in the concrete industry may lead to greater use of this alternative.

More use of higher strength steels (above grade S355) is likely to be seen in future. Several European countries are making significant use of S420 and S460 grades, and it has been shown that they can be more economic than S355 in some circumstances.

The designer should always try to be aware of new techniques in fabrication and construction and should be receptive to alternative proposals offered by contractors.

5 DETAILED DESIGN: MAIN BEAMS

The detailed design stage confirms or refines the outline design produced in the initial design stage. It is essentially a checking process, applying a complete range of loading conditions to a mathematical model to generate calculated forces and stresses at critical locations in the structure. These forces and stresses are then checked to ensure that they comply with the good practice expressed in the Code. The detail of the checking process is sufficiently thorough to enable working drawings to be prepared, in conjunction with a specification for workmanship and materials, and the bridge to be constructed.

5.1 Global analysis

A global analysis is required to establish the maximum forces and moments at the critical parts of the bridge, under the variety of possible loading conditions. Local analysis of the deck slab is usually treated separately from the global analysis; this is described further in Section 7.

It is now common practice to use a computer analysis, and this facility is assumed to be available to the designer. Software packages are available over a wide range of sophistication and capability, and the selection of program will usually depend on the designer's in-house computing facilities. For a structure as fundamentally simple as a beam-and-slab bridge, quite simple programs will usually suffice.

For a simple span, the central bending moment on the whole bridge is of course easily calculated manually. However, the proportion of the moment carried by each beam depends on the relative stiffnesses of beams and slab, so a computer model analysis is of benefit even for such structures.

5.1.1 Types of computer model

The basis of most commonly used computer models is the grillage analogy, as described by West[20] and Hambly[21]. In this model, the structure is idealised as a number of longitudinal and transverse beam elements in a single plane, rigidly interconnected at nodes. Transverse beams may be orthogonal or skewed with respect to the longitudinal beams.

Each beam element represents either a composite section (e.g. main girder with associated slab) or a width of slab (e.g. a transverse element may represent a width of slab equal to the spacing of the transverse elements). Examples of typical grillages are shown in Figure 5.1 .

A refinement of the basic grillage model is the grid-and-slab model, which adds plate bending elements to the grillage of beams. This simplifies the representation of the transverse action of the slab and permits the element forces to be output as unit forces per width of slab. It would also enable local effects (wheel loadings) to be modelled at the same time as global effects, if a sufficiently fine mesh were used, although this would make a very significant increase in the size of the computer model; local effects are therefore usually assessed separately (see Section 7.2).

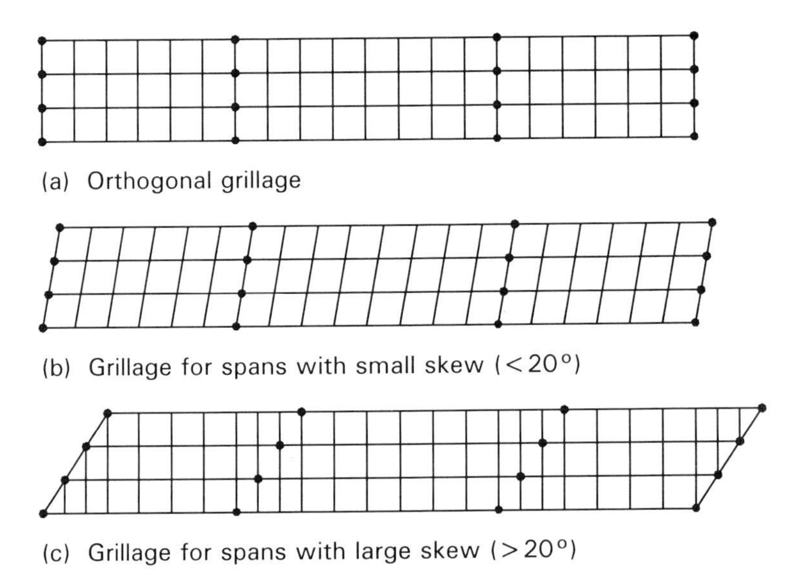

Figure 5.1 *Typical grillages*

The increasing availability of analytical software may lead to wider use of the more sophisticated models, though at present the use of simple grillages is strongly predominant.

5.1.2 Parameters for global analysis

It is usual to choose the longitudinal beam elements so that they coincide with the main girders. An edge beam may be provided at the edges of the slab if the cantilevers need to be modelled, but that is not usually necessary. Because the transverse beams do not represent discrete elements, the spacing can be chosen by the designer to facilitate the analysis. Generally, the spacing should not exceed about 1/8 of the span. Uniform node spacing should be chosen in each direction where possible.

For skew spans where the transverse reinforcement is parallel to the abutments (i.e. for skew angles less than 20° - see further comment in Section 7.2), the transverse members should follow the line of the reinforcement. For larger skews, the grillage should be orthogonal and the mesh chosen to align with supports (see Figure 5.1).

Gross section properties should be used in the global analysis (Clauses 3/7.2, 5/5.1.1 and 5/6.1.4.1). Section properties for longitudinal beams should assume that the fully effective slab is shared between the beams. Many designers consider it adequate to use only short-term concrete properties for the global analysis, rather than deal with two sets of properties; this results in a very slightly reduced design moment in cracked sections over supports and correspondingly higher midspan moments. Long- and short-term load effects should nevertheless be determined separately, because they should be applied separately to long- and short-term section properties in the stress analysis of sections.

Section properties for transverse beam elements representing transverse bracing or cross girders should be determined on the basis of both the bending stiffness and the shear stiffness of the bracing acting with deck slab.

Section properties for transverse beam elements representing the slab alone should use a width equal to the element spacing. Torsional stiffness of the slab should be divided equally between the transverse and longitudinal beams; use $bt^3/6$ in each direction, where b is the width of slab appropriate to the element concerned.

5.1.3 Choice of cracked or uncracked analysis

For the loads applied to a continuous composite structure at ULS, global analysis may be carried out assuming initially that the concrete slab is uncracked over internal supports; up to 10% of the support moments may then be distributed to the span. Alternatively, and more usually, the concrete may be assumed to be cracked for a length of 15% of the span on each side of an internal support (Clause 5/6.1.4). Similarly the slab may be assumed initially to be uncracked for SLS and fatigue analysis, but if the concrete stress exceeds $0.1 f_{cu}$, either a new analysis assuming cracked concrete at the supports is required or the midspan moments must be increased without corresponding reduction in hogging moment (Clause 5/5.1.1).

Generally, it is advisable for the global analysis for both SLS and ULS to assume from the outset that the concrete is cracked adjacent to internal supports for about 15% of the span. No further redistribution should then be made.

[Note that even with this latter analysis, the BSI version of Part 5 allowed a further redistribution of moments at ULS with compact sections (Clause 5/6.1.4.2), but this has been deleted in BD 16/82 and the combined version of Part 5, as it relates to plastic moment distribution, which is not permitted.]

Cracked section properties should include the effective area of the reinforcement over the full width of slab acting with the steel girder. This effective area should take account of the slightly lower modulus of elasticity of the reinforcement.

5.1.4 Analysis of staged construction

It is usual for the deck slab of composite bridges to be concreted in stages, and for the steel girders to be unpropped between supports during this process. Part of the load is thus carried on the steel beam sections alone, part by the composite sections. A number of separate analyses may be required, one representing each different stage that occurs. This series of analyses will follow the concreting sequence and will take account of the distribution of the weight of wet concrete, particularly that of the cantilevers (Clauses 5/5.1 and 5/12.1). It will be a series of partially composite structures.

Typically there are about twice as many stages as spans, because concrete is placed successively in each of the midspan regions, followed by the remaining regions over each support. Where the cantilevers are concreted at a different stage from the main width of slab, this must be taken into account in the analyses.

The total deflections under unfactored dead and superimposed loads should be calculated to enable the beams to be pre-cambered. This information should be produced by the designer and included on the drawings.

5.1.5 Influence surfaces

Selection of the worst position for application of the HB vehicle can be greatly aided if the grillage analysis program generates influence surface data. The surface can often be readily displayed by use of graphics software.

Influence surfaces can be very useful with highly skewed bridges. The acute corner bearings can be subject to negative reaction under certain loading; influence surfaces can determine where this occurs.

5.2 Load effects and combinations

The loadings to be applied to the bridge are all specified in Part 2 (or rather the modified Part 2, issued in the Appendix to BD 37/88 - see Section 2.7), except for the standard fatigue vehicle, which is specified in Part 10. Table 1 of Part 2 specifies the appropriate partial factors to be applied to each of these loads, according to the combinations in which loadings occur.

Because many different load factors and combinations are involved in the assessment of design loads at several principal sections, it is usual for each load to be analysed separately and without load factors. Combination of appropriate factored loadcases is then either performed manually, usually by presentation in tabular form, or, if the program allows, as a separate presentation of combined factored forces. Because so many separate loadcases and factors are used to build up total figures, the designer is advised to include routine checks (such as totalling reactions) and to use tabular presentation of results to avoid errors. The graphical displays and print-outs provided now by analysis and spreadsheet software can also be recommended for checking results.

The object of the analysis is to arrive at design load effects (moments, shear forces and stresses) for the various elements of the structure. The most severe selection of loadings and combinations needs to be determined for each critical element. The main design load effects that are to be calculated include the following:

- Maximum moment with co-existent shear in the most heavily loaded main girder: at midspan, over intermediate support and at splice positions.
- Maximum shear with co-existent moment in the most heavily loaded main girder: at supports and at splices.
- Maximum forces in transverse bracing at supports (and in participating intermediate bracing).
- Maximum and minimum reactions at bearings.
- Transverse slab moments (to be combined with local slab moments for design of slab reinforcement).
- Range of forces and moments due to fatigue loading (for shear connectors and any other welded details that need to be checked).

Displacements and rotations at bearings will also need to be calculated. Displacements at midspan may need to be checked if clearance is critical. Stiffnesses of the girders during construction may need to be determined, for design of bracing and restraint systems.

The total deflections under dead and superimposed loads should be calculated so that the designer can indicate the dead load deflections on the drawings.

Selection of the most heavily loaded girder can usually be made by inspection, as can selection of the more heavily loaded intermediate supports. Influence lines can be used to identify appropriate loaded lengths for the maximum effects. If cross sections vary within spans, or spans are unequal, more cases will need to be analysed to determine load effects at the points of change or in each span.

It is usually found that the specified Combination 1 (see Clause 2/4.4.1) governs most or all of the structure. Some parts, notably top flanges, are governed by construction conditions, Combination 2 or 3. For spans over about 50 m, Combination 2, including wind load, may determine design of transverse bracing and bearing restraint.

In the preparation of a bridge bearing schedule (Clause 9.1/A.1), it should be made clear that the tabulated forces are load effects, i.e. they include both γ_{fL} and γ_{f3}, and whether the effects are at ULS or SLS.

5.2.1 Temperature effects and shrinkage modified by creep

The effects of differential temperature and shrinkage modified by creep need to be considered at SLS (Clauses 3/9.2.3 and 5/5.4). Similar effects need to be considered at ULS, except when the section is compact throughout the span and the member is stable against lateral torsional buckling [λ_{LT} less than about 30, see 3/9.2.1.3(d)].

Differential temperature (temperature gradient through the depth of the beams) and shrinkage give rise to 'internal' strains and are not like the externally applied dead and live loads. The effects due to differential temperature and shrinkage modified by creep need to be calculated in two parts. The first is an internal stress distribution, assuming that the beam is free to adopt any curvature that this produces (primary effects). The second is a set of moments and shears necessary to achieve continuity over a number of fixed supports (secondary effects). See Figure 5.2.

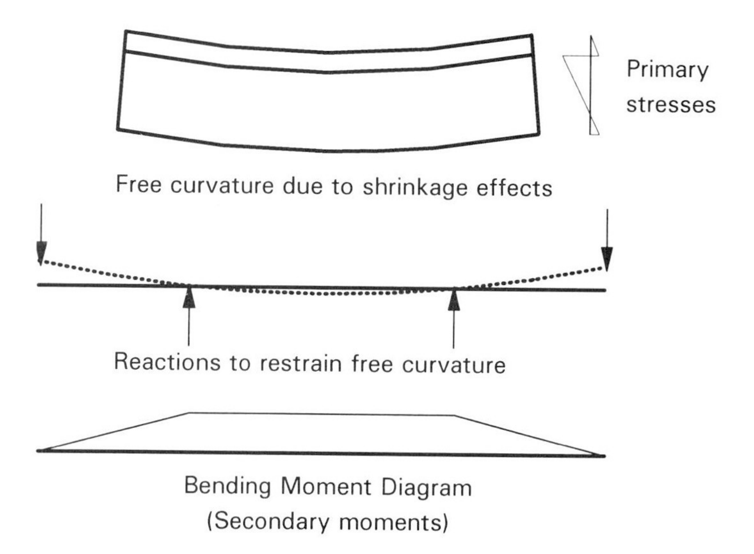

Figure 5.2 *Primary and secondary effects due to shrinkage*

Part 3 deals separately with these primary and secondary effects (see Clause 3/9.9.7, which should be applied to beams with longitudinal stiffeners as well as those without, although Part 3 does not explicitly say so). The primary effects are calculated assuming uncracked sections, the secondary effects are calculated allowing for concrete cracked in tension. Both primary and secondary effects need to be considered at SLS but only the secondary effects need to be considered at ULS (and subject to the exception mentioned above). The argument seems to be that the internal primary stresses can redistribute locally at ULS but the secondary effects (which depend on the behaviour of the whole length of the beam) do not, except when there is plastic redistribution with a compact section.

The effects for differential temperature and shrinkage are specified in Clauses 2/5.4.5 and 5/5.4.3 respectively. Partial factors γ_{fL} are given for differential temperature in Clause 2/5.4.8.4 and for shrinkage in Clause 5/4.1.2.

The variation of the 'effective bridge temperature' depends on the range of ambient conditions for the bridge site and on the form of construction. Isotherms for the UK are given in Clause 2/5.4. The resulting expansion or contraction can give rise to eccentric reactions at movement bearings and to restraint forces in integral bridges (see Section 5.2.4).

All temperature effects are considered to be in Combination 3. Shrinkage effects should be considered in all combinations, but only if the effects are adverse.

5.2.2 Moment gradient adjacent to intermediate supports

At intermediate supports, the maximum effects (moment and shear) occur at (or, in the case of shear, just to one side of) the line of support. The moments and shears decrease quickly away from the supports into the span. Strength requirements depend mainly on these maximum effects but, where the strength is limited by buckling of the bottom flange, the rate of decrease of moment to the next effective restraint can have a significant effect. In such cases, the variation in moment and shear given by the global analysis should be noted.

5.2.3 Co-existent effects in sagging regions

When selecting the worst effects for design, it should be noted that bending/shear interaction for beams without longitudinal stiffeners is checked for the worst effects over the length of a web panel (between transverse stiffeners), not simply for the effects at a single cross section. In midspan sagging regions there might be few transverse web stiffeners and the web panels will be quite long. The maximum moment and maximum co-existent shear would then occur some distance apart, in some cases at the opposite ends of the panel. Care should be taken to extract the correct load effects from the global analysis for verifying the strength of the beam.

5.2.4 Effects in integral bridges

In an integral bridge, where the ends of the bridge are continuous with the supporting structure, the bending moment distribution is modified (from that with a pinned end support) because of the partial rotational restraint that is provided. Also, temperature variation (of the mean bridge temperature) gives rise to axial forces in the main beams. Guidance on the evaluation of these effects is given in *Integral steel bridges: Design guidance*[13]. It is noted there that the axial effects are not likely to be large and, being a Combination 3 loadcase (with smaller partial factors on live load), are not likely to govern the design of the beams. However, in some circumstances the axial force might change the section classification to non-compact (because a greater depth of web would be in compression), which might have a significant effect on the design.

5.3 Design of beams

The main longitudinal beams must be designed to provide adequate strength in bending and shear to resist the combined effects of global bending, local effects such as direct wheel loading or compression over bearings, and structural participation with bracing systems.

Part 3 provides two sets of Clauses for evaluating the strength of beams, one for beams without longitudinal stiffeners (3/9.9) and one for beams with longitudinal stiffeners (3/9.10 and 3/9.11). The first evaluates the strength of the beam cross section as a whole, and the second evaluates the strength in terms of its elements (flange panels and web panels, between stiffeners). Guidance on the two different treatments is given separately in Sections 5.4 and 5.5 below.

Additionally, for beams without longitudinal stiffeners, Part 3 (and also Part 5) recognises a difference between sections that it terms 'compact', where the plastic moment resistance of the section can be developed, and those that it terms 'non-compact', where local buckling of the web or flange limits its resistance to that at first yield in an extreme fibre. Sections 5.3.3 to 5.3.5 below explain the differences between the two classes and the implications for design.

5.3.1 Requirements at ultimate and serviceability limit states

Part 3 requires that the resistances at all sections of the beams must satisfy the requirements at ULS. In determining bending resistance at ULS, the effects of shear lag should be ignored.

Clause 3/9.2.3 states that the requirements at SLS need only be checked for the structural steel elements when one (or more) of the following conditions arise:

(i) Shear lag is significant; this arises when the effective breadth ratio is less than a restricting value given by Clause 3/9.2.3.1(a). Note that, for composite construction, the effective breadth ratio is modified by Clause 5/5.2.3. The restricting ratio is likely to be just over 0.7 for typical composite bridges, and this will only be exceeded for widely spaced main girders - girder spacing at least 10% of the main span, for example (see Clause 3/8.2 for tables of values).

(ii) Tension flange stresses have been redistributed at ULS in accordance with Clause 3/9.5.5. This redistribution is not usually employed in this form of construction, because either the compressive stresses govern (hogging regions) or little advantage is gained (tension flanges in sagging regions are much heavier than webs).

(iii) An unsymmetric beam is designed as a compact section at ULS (all composite sections are normally unsymmetric).

For composite beams, Part 5 requires that the stresses in reinforcement and concrete should be checked at SLS, taking account of any co-existent stresses due to local bending of the slab (Clauses 5/5.2.4 and 5/5.2.6). Calculation of stresses at SLS should allow for the effects of shear lag (Clauses 3/8.2 and 5/5.2.3). Crack widths in tensile regions of the slab should be checked at SLS.

For further discussion of the design of the deck slab, see Section 7.

5.3.2 Construction in stages

Composite bridges are usually built in stages: the steel girders are erected first; the concrete slab is cast over part of the bridge; the slab is cast over successive parts until complete; the surfacing and the parapets are added. At each stage, the structure is different, and the distribution of bending moments (in a continuous bridge) and the interaction between steel and concrete elements are different. Additionally, the response of the completed structure to live load is different again, because the concrete is stiffer in response to short-term loading.

Consequently, in most cases it is necessary to determine moments and shear at the various different cross sections as summations of a number of values. This summation is necessary even for beams that are deemed to be compact, although in those cases some of the load effects can be neglected, because it is assumed that they can redistribute within the structure at ULS.

5.3.3 Definition of compact and non-compact sections

Compact sections are those where 'the web and compression flange possess sufficient stiffness to enable full plasticity and adequate rotation to be developed without the loss of strength due to local buckling' (Clause 5/6.2.2.1). Limits are given in Part 3 for breadth/thickness ratios for which this requirement is deemed to be true (Clause 3/9.3.7).

Part 3 does not explicitly define "non-compact" sections, it merely refers to them as the alternative to compact sections. There is, nevertheless, a limit to the outstands of compression flanges that applies to non-compact sections (Clause 3/9.3.2.1); this limit allows a flange to develop its full yield strength, although it cannot sustain plastic strain like the flange of a compact section.

5.3.4 Design implications for compact sections

When a plastic moment is developed in a beam, the stress distribution through the section takes the form of rectangular blocks rather than the usual 'triangular' distribution (Figure 5.3). The amount by which the plastic moment exceeds the elastic moment when yield is just reached depends on the relative proportions of web and flange; the ratio between the two moments is commonly known as the shape factor. For a compact composite Universal Beam in sagging, the shape factor is typically between 1.05 and 1.20.

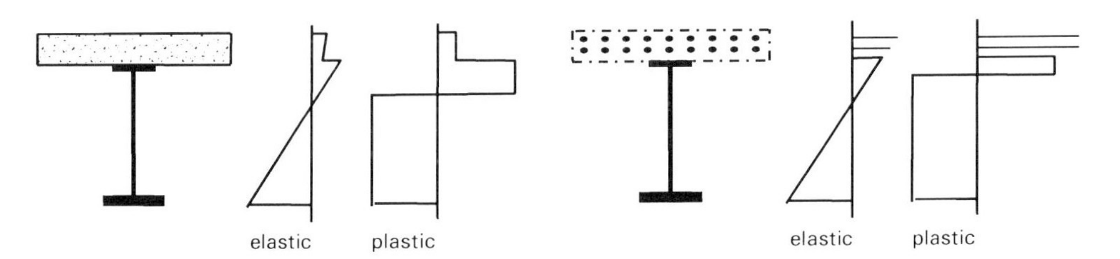

Figure 5.3 *Stress distributions in composite beams*

Because the bending resistance of compact sections at ULS is determined by redistribution of the stresses at any cross section, the design requirement for a compact section is simply that the plastic moment capacity is adequate to carry the total moment acting at that cross section at ULS (Clause 5/6.2.2). It is not necessary to check the summation of the separate ULS stress distributions for each stage of construction.

Although the strength of compact sections is determined from a plastic stress distribution, the distribution of bending moments should still be determined by elastic analysis. Indeed, although the BSI version of Part 5 permitted the use of plastic analysis for members that are compact throughout their length (provided that special considerations are made), BD16/82 and the combined version of Part 5, which are the rules normally used, permit only elastic analysis for moment distribution (see Section 5.1.3 above).

The mobilisation of plastic moment capacity must result in plastic bending (or a rotation of the beam axis over the affected length) and this must be accompanied by consistent rotational displacements at other parts of the beam. For example, at the stage when plastic moments have developed at midspan, a considerable length of the bottom flange in the span will have undergone plastic strain, and a significant rotation will also have occurred at or adjacent to the internal supports. (This must be so, even though moment distribution has been determined by elastic analysis.) A conservative view would therefore be to design any beam section as compact only if *all* the beam sections throughout its length are compact. However, investigation of a wide range of possible proportions of spans and composite sections[22] has shown that because the design loading for the two regions (hogging and sagging) differs, each region may be designed on its own merits, and compact moment resistance may be used in midspan regions when non-compact resistance is used at intermediate supports.

Indeed, in most cases of composite beam and slab construction the hogging moment regions over intermediate supports are unlikely to be compact (as defined by 3/9.3.7): the web would need to be much thicker than needed for shear capacity to achieve compact classification.

If the beam is compact throughout the span, and the member is not too slender (see comment in Section 5.2.1), the effects of differential temperature, creep and settlement may be neglected at ULS [see Clause 3/9.2.1.3(iv)]. This allowance appears to be based on the presumption that the secondary moments (due to temperature gradient and shrinkage) can redistribute in such cases.

Unsymmetric beams designed to plastic moment capacity will reach yield in one flange before the other and before the full moment resistance is developed. Consequently, the stress in the more highly stressed flange must be checked elastically (Clause 3/9.9.8) at SLS to ensure that yield is not exceeded at that state (see comment on summation of stresses in Section 5.3.5). The effects of temperature gradient and shrinkage must then be included.

To achieve plastic moment in a composite section, the shear connectors must be capable of transferring an appropriate shear between the beam and the slab. This is discussed further in Section 7.5.

It should be noted that steel beams that are compact when acting compositely with the slab may not be compact when acting alone during construction. In such a case, the checks for the construction condition must be made on the basis of non-compact sections.

5.3.5 Design implications for non-compact sections

The design strength of non-compact beam sections is determined by essentially elastic stress distributions and limiting stresses. In most cases, therefore, checks need only be made at ULS. Checks at SLS are required in only a few circumstances (Clause 3/9.2.3.1).

A non-compact composite section must provide sufficient moment resistance for the total moment acting at that section (determined as a summation, as explained in section 5.3.2 above). However, because a non-compact section is not able to redistribute stresses from the compression flange to the web, a check must also be made on the summation of the elastically determined stresses. The total stresses and strains in the fibres of a composite beam where the deck slab is added in stages are determined as the summations of the distributions for each stage (and similarly for short- and long-term loads), as shown diagrammatically in Figure 5.4. The position of zero stress will therefore not necessarily correspond with any particular neutral axis level.

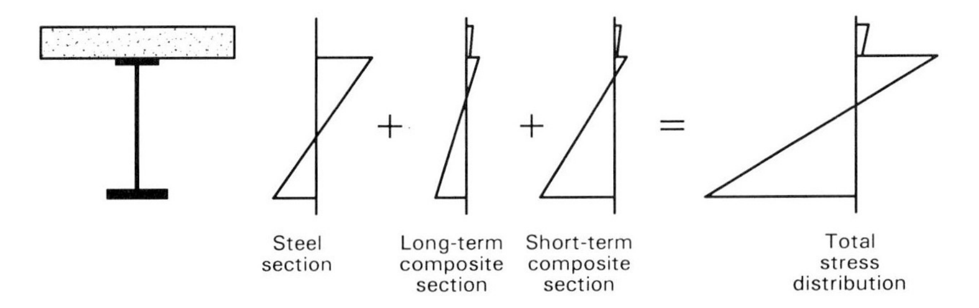

Figure 5.4 *Summation of stresses for staged construction*

Where SLS must be checked as well as ULS, the stress distributions for SLS and ULS must be calculated separately, each using its appropriate set of partial factors for the various loads.

5.3.6 Forms of stiffening to beams

Economic design of longer spans, usually minimising the cross-sectional area of the steel girder, results in relatively thin webs. Slender webs will have limited shear capacity, especially when combined with high flexural stresses. It is customary to provide some stiffening to a thin web to enable it to carry higher loads, though an economic balance has to be struck between the cost of stiffening and the cost of increasing the thickness of the web.

Web stiffeners

The most common form of web stiffening is the transverse (nominally vertical) stiffener, welded on one face of the web. This increases the shear capacity of thin webs considerably by restricting shear buckling to shorter panels and by allowing so-called tension field action to develop across the diagonals in a truss-like pattern. It is typically used when the web depth/thickness (d/t) exceeds about 50. A web d/t up to as much as 200 may be used in a web stiffened only by transverse stiffeners.

At the higher web d/t values, the part of the web that is in compression may be unable to carry sufficient combined shear and direct stress, because of buckling of the plate between transverse stiffeners. To restrain this buckling and carry greater load, longitudinal stiffeners can be welded onto the web. This increases fabrication costs, so these stiffeners are typically used only in the compression region over intermediate supports (particularly with haunched girders) for spans over about 50 m.

The two forms of web stiffening, transverse and longitudinal, lead to different methods of strength evaluation in Part 3. When there are no longitudinal stiffeners, the strength in bending is determined as an overall moment capacity; the strength in shear is the capacity of the full depth of web. Where longitudinal stiffeners are provided, a check for buckling and yield is required separately for the flanges, for each panel bounded by stiffeners and for each effective stiffener section (the stiffener with its associated width of web).

Beams of variable depth cannot be designed using the 'simple' rules for unstiffened beams. Beams with curved soffits must be treated in the same way as longitudinally stiffened beams (see 3/9.9.1.1), and beams with non-parallel flanges must be designed for shear as a longitudinally stiffened beam (see 3/9.9.2.1), although moment resistance can still be evaluated in accordance with Clause 3/9.9.1.

Flange stiffeners

Longitudinal stiffening to a compression flange is often used in box girders but it is not usual with beam-and-slab construction.

Types of longitudinal stiffening covered in this publication

In this guide the only form of longitudinal stiffening considered is that provided on the web.

5.3.7 Bracing to main beams

Bracing is required to restrain the primary members (the longitudinal beams) and to provide a load path for transferring horizontal forces (transverse to the main beams), particularly at support positions.

All beams are required (Clause 3/9.12.5.1) to be provided with torsional restraints at supports, because they will have been designed to resist lateral torsional buckling (see Section 5.4.2). These torsional restraints must also be capable of transferring lateral loads such as wind load and skidding loads down to the level of the bottom flanges (from where they will be transferred to the bearings). The different forms of bracing at supports are discussed in Section 6.2.

With simple spans, no intermediate bracing is normally needed in the completed structure, but some bracing is usually required to stabilise the top flanges during construction (see Section 6.3) and to provide a load path for wind loads before the deck slab is complete.

With continuous spans, beams are usually braced adjacent to intermediate supports to provide restraint to the compression flange (again, see Section 5.4.2), as well as in midspan regions for construction considerations. Examples of intermediate bracing arrangements are given in Section 6.3.

Bracing across the width of a multiple girder bridge would act as a stiff transverse element and thus help to distribute concentrated loads. However, because a continuous transverse bracing system will participate in distributing all loads transversely, the bracing members will be subject to load reversal; if not detailed properly they would be prone to fatigue. The normal arrangement with multiple beam bridges is therefore to brace the beams in pairs, with no bracing in the bays between the pairs. This makes the transverse bracing 'non-participating' in load distribution. To facilitate such pairing, designers are encouraged to use an even number of main beams.

5.4 Beams without longitudinal stiffeners

Bending and shear resistances of a beam without longitudinal stiffeners are assessed separately and the effects of interaction between the two are taken into account by expressions that define a limiting envelope for the combined load effects. Naturally, the combination of bending and shear is greatest at internal supports, though it is still necessary to check other locations, such as the weaker side of a splice, where the moment may be close to the bending resistance of the section.

5.4.1 Bending resistance

The bending resistance M_D of a beam without longitudinal stiffeners is expressed in terms of the limiting moment of resistance M_R divided by partial factors γ_m and γ_{f3} (Clause 3/9.9.1.2). The value of M_R is determined by the strength of the cross section M_{ult} multiplied by a reduction factor (where appropriate) to allow for lateral torsional buckling (Clause 3/9.8 and Figure 11).

Lateral torsional buckling (LTB) is a well-known phenomenon in slender beams that are free to buckle between positions of restraint, and the behaviour is taken into account by use of a 'buckling curve' such as that in Figure 11. Figure 11 gives the value of the ratio M_R to M_{ult} for a modified value of the slenderness parameter λ_{LT}. See further comment in Section 5.4.2 about the calculation of the modified slenderness parameter.

For non-compact sections, the value of M_{ult} is the strength at first yield in an extreme fibre of the cross section (determined from the effective section properties, with no allowance for shear lag but reduced for holes and slender webs, Clause 3/9.4.2). For compact sections, the value of M_{ult} is the plastic moment capacity of the effective section, M_{pe}.

The use of this approach for compact sections, using a plastic section property and a curve that is essentially dependent on elastic buckling, is difficult to envisage in physical terms, but calibration against experimental tests confirms that it is a reasonable evaluation.

For composite sections that are typical in highway bridges, the elastic bending resistance is not normally governed by the strength of either the concrete or the reinforcement, but the plastic moment capacity does depend on the strength of those elements. (And, as noted in Section 5.4.2, the value of M_{pe} is needed even for non-compact sections.) See Section 7.3.1 for comment on values of design strength for concrete and reinforcement.

5.4.2 Effective length and slenderness

The determination of the ratio M_R/M_{ult} involves the calculation of effective length for lateral torsional buckling (ℓ_e) and slenderness parameter (λ_{LT}). These parameters depend on bracing arrangements and moment variation along the beam (Clauses 3/9.6 and 3/9.7).

During construction, when there is no slab, the girders can buckle in a true lateral-torsional mode. However, once the slab is complete, only the compression flange in the hogging regions can buckle laterally; this is actually a distortional buckling phenomenon but for simplicity the code maintains a single form of treatment. The simplification is conservative.

Part 3 covers four different ways in which a compression flange can be stabilised against buckling laterally out of its plane.

Continuous direct restraint

The first and most direct form of restraint is when the flange is connected directly to a deck (i.e. top flange of a single span composite beam and slab bridge and the midspan portion of a continuous composite bridge). Then the effective length is zero (Clause 3/9.6.4.2.1) and M_R simply equals M_{ult}.

Discrete intermediate lateral and torsional restraints

The second form of restraint is more often seen during construction, when the compression flanges of the bare steel **I** beams are typically restrained at discrete positions by means of lateral or torsional restraints.

Where there is plan bracing to the top flange, this provides discrete lateral restraint that is so stiff that it is fully effective, and then the buckling effective length is equal to the spacing of the restraints (Clause 3/9.6.4.1.1.1). Similarly, a direct connection of the flange to another structure (say a parallel length of completed deck) would be fully effective.

However, plan bracing to the compression flange is not essential if stiff torsional restraints are provided between two beams at discrete intervals. This effectively creates a 'torsional spring restraint' that modifies the buckling mode between support positions. The effective length with this form of restraint is given by Clause 3/9.6.4.1.2. If the torsional stiffness provided by the action of the pair of beams is sufficiently high, the effective length can be restricted to about one-third of the span, but any further restriction cannot practically be achieved (see the 'distributed restraints' curve in 3/Figure 8).

Note that, for longer spans, plan bracing may be needed to carry wind loads during construction.

In service, the bottom flanges adjacent to intermediate supports are often restrained at discrete positions by cross bracing acting in conjunction with the deck slab (see Section 6.3.1). This again provides fully effective discrete restraints.

Discrete U-frame restraints

Where ladder-beam construction is used, the compression flanges of a continuous composite bridge adjacent to intermediate supports are restrained at discrete positions not by cross bracing but by the stiff frames formed by the cross-beams (with deck slab) and the web stiffeners (see the bottom diagram in Figure 4.1). These are referred to as discrete U-frames. The stiffnesses of such restraints are evaluated according to Clause 3/9.6.4.1.3 and the effective length is determined according to Clause 3/9.6.4.1.1.2.

Continuous U-frame restraint

If the compression flange adjacent to an intermediate support is not provided with any discrete restraint, it may still receive some restraint from the U-frame action of the slab with the web plate. This is termed continuous U-frame restraint and the effective length is determined (Clause 3/9.6.4.2.2) by considering the restraint as a series of U-frames, each of unit length, then determining stiffness and effective length in the same way as for discrete U-frames.

Half-wavelength of buckling

For restraints that are fully effective (i.e. they prevent displacement and create nodes in the mode of buckling), the effective length and the half-wavelength of buckling are the same. But whilst restraints that are less than fully effective can increase the buckling load, they may not entirely prevent displacement at the restraint positions; the half-wavelength of buckling is then greater than the effective length. This has implications for the determination of the moment of resistance.

Use of slenderness to determine limiting moment of resistance

Once the effective length has been determined, the slenderness is calculated according to Clause 3/9.7. The expression for λ_{LT} includes a parameter η that allows for any moment variation over the half-wavelength of buckling. In midspan regions, where there is little variation, the parameter is close to 1.0; at intermediate supports it can make a significant reduction in the value of λ_{LT}. In either case, η may always be taken conservatively as 1.0.

As mentioned in Section 5.4.1, the determination of M_R involves the use of the modified slenderness parameter and a buckling curve. The modification to the value of the slenderness parameter λ_{LT} is the multiplication by two ratios. The first ratio is the value $\sqrt{(\sigma_{yc}/355)}$. This allows for the actual strength of the material (Figure 11 is drawn for the specific value of 355 N/mm^2 for the yield strength). The second ratio is the value $\sqrt{M_{ult} / M_{pe}}$. For compact sections, this value is unity. For non-compact sections, the value is less than unity: although it may seem surprising that this requires the calculation of M_{pe} for a non-compact section, the modification is needed because the buckling curve in Figure 11 has been calibrated, for both compact and non-compact sections, against the value of M_{pe}.

The buckling curve in Figure 11 determines the moment resistance on the basis of the slenderness and an assumed initial imperfection. The assumed imperfection is taken to be related to the half-wavelength of buckling, rather than the effective length, and it includes an allowance for residual stresses. Two Figures are given in Part 3, one for welded members and one for non-welded or stress-relieved members (the allowance for residual stress is different). Also, in each Figure, three curves are given, for different values of the ratio effective length / half-wavelength of buckling (ℓ_e/ℓ_w). The ratio 1.0 is the 'ordinary' case, for fully effective restraints.

For a more detailed explanation of the determination of slenderness and limiting moment of resistance, see *Commentary on BS 5400-3:2000, Code of practice for the design of steel bridges*[3].

Design of restraint system

The requirements for the design of the restraints themselves are discussed in Section 6.

5.4.3 Summation of stresses

As mentioned in Section 5.3, the stresses through a non-compact section that is constructed in stages have to be determined by summation of several stress distributions. (This applies also to compact beams treated as non-compact at SLS.) The resulting stress variation is therefore non-linear and discontinuous, and could not be derived from any single application of loads to a single section.

For these cases, verification of adequate bending resistance at ULS depends on two requirements: first, that the total moment (the sum of all the stage moments) should not exceed M_D (see Section 5.4.1 for comment on bending resistance); and second, that the total accumulated stress at any fibre should not exceed the factored yield strength (Clause 3/9.9.5.4). This second requirement also applies to the SLS check on compact sections.

5.4.4 Shear resistance

An unstiffened slender web is unable to develop full shear yield resistance, because its capacity is limited by buckling. However, when the web is provided with transverse stiffeners, the buckling resistance is increased. The increase arises firstly from the constraint of the rectangular panel and secondly because some of the shear is carried by tension field action. The mechanism for the latter is shown in Figure 5.5, which also shows how the magnitude of the tension field component is enhanced where the flanges are stiff and plastic hinges develop in them.

The design shear resistance V_D (Clause 3/9.9.2.2) is the shear capacity of the web panel, including the contribution to the tension field action from the flanges.

Intermediate web stiffeners are an effective means of increasing the limiting shear stress when the web depth/thickness is in excess of about 75. Typically, intermediate web stiffeners are provided at a spacing of between 1.0 d and 2.0 d.

The same rules for calculating shear resistance apply to webs in both compact and non-compact sections. However, the webs in compact sections are less slender; by definition the depth in compression is limited and the overall depth will rarely be sufficient to require intermediate transverse stiffeners.

5.4.5 Bending-shear interaction

Under combined bending and shear, a bending resistance M_f equal to that provided by the flanges alone, ignoring any contribution from the web, can be sustained at the same time as the shear resistance V_R if contribution from the flanges is ignored. Further, it has also been shown that the full bending resistance M_D can be developed if the shear is not more than half of V_R, and that full shear resistance V_D can be developed if the bending is not more than half of M_f. These limits to the interaction are expressed in Clause 3/9.9.3.1 and are shown graphically in Figure 5.6.

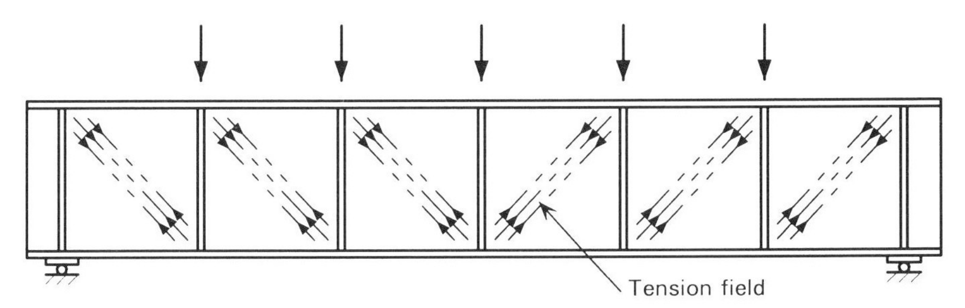

(a) Tension field in a stiffened web (analogous to N truss)

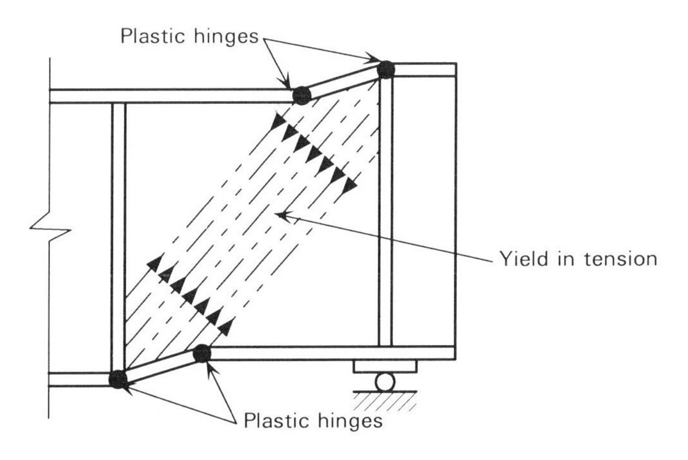

(b) Development of plastic hinges

Figure 5.5 *Tension field action in webs*

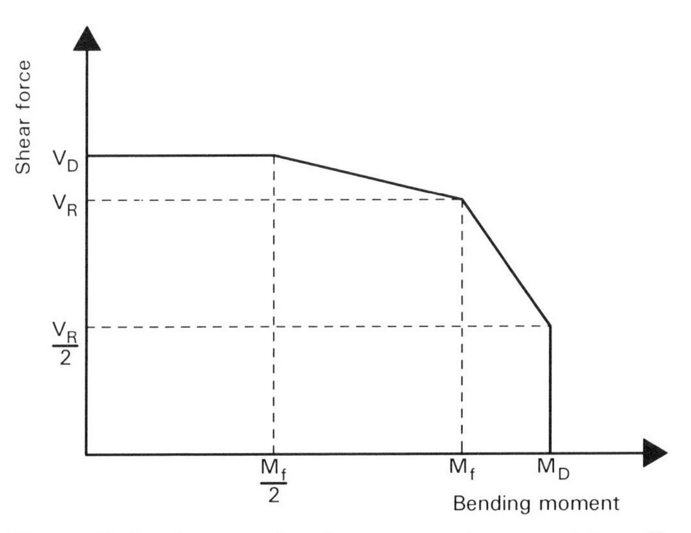

Figure 5.6 *Interaction between shear and bending resistance*

Bending-shear interaction is checked for the worst moment and worst shear anywhere within a panel length (i.e. between intermediate web stiffeners), rather than at a single section. This is only slightly conservative for panels adjacent to internal supports but seems rather onerous for midspan regions of beams of compact section, where the panel could extend between stiffeners at the points of contraflexure (midspan moment is then combined with shear at the zero moment position).

For beams constructed in stages, the moment acting on the section should be taken as the total moment for sections designed as compact, but for sections designed as non-compact an effective bending moment must be derived for use in the interaction formulae. This is obtained by multiplying the extreme fibre total stress by the modulus for that fibre in the section that is appropriate to the stage of construction being checked (see Clause 3/9.9.5.5). The designer should take care to ensure that the fibre for which total design stress is used to determine the equivalent bending moment is the same fibre that determines the bending resistance.

5.4.6 Adequacy during construction

Stability of the compression flange must be achieved during construction as well as in service. In sagging regions the resistance of a composite beam at ULS is usually governed by the tensile yield of the bottom flange (although in some circumstances the compression in the slab may govern). During concreting, with the weight of wet concrete carried only by the steel beams, lateral torsional buckling and the stability of the top (compression) flange may well govern the design. Adequate bracing to the top flange must be provided for this condition, although it may be temporary and can be removed after concreting. Cross bracing (between a pair of beams) if often sufficient for this purpose, although plan bracing to the flanges may be required for carrying lateral wind loads in longer spans (but it adds to the complexity of fabrication and erection, and should be avoided where possible).

5.5 Beams with longitudinal stiffeners

If a web without longitudinal stiffeners is very deep ($d/t > 200$) it contributes little or nothing to the bending resistance and has a low limiting shear stress. Such deep webs are common in haunched girders and girders with curved soffits. Longitudinal web stiffeners are then provided locally to the supports, which increases both bending and shear resistance. However, the beam web must then be subject to a more detailed check, on a panel by panel basis (Clause 3/9.11).

As for unstiffened beams, the resistance at all sections must satisfy the requirements of the ULS. Requirements for SLS need only be satisfied in the deck slab and in the special circumstances mentioned in Section 5.3.1 above.

Three separate but related aspects of the strength of the beam section must be checked for longitudinally stiffened beams, and these rules apply also to beams with curved soffits or to the shear resistance of beams with non-parallel flanges.

5.5.1 Bending resistance

The requirements for bending resistance are expressed in terms of limiting stresses in the extreme fibres of the flanges. The summation of stresses for any fibre (calculated on the effective sections at each stage) must not exceed limiting values in compression or tension (Clause 3/9.10.1). The limiting compressive stress depends on the yield strength and on lateral torsional buckling, as for bending resistance of beams without longitudinal stiffeners; the value is calculated by dividing the value of M_R (evaluated in accordance with Clauses 3/9.6, 9.7 and 9.8) by the elastic section modulus Z_{xc} and by the partial factors $\gamma_m\gamma_{f3}$. In tension, the limiting stress is simply the yield stress of the flange σ_{yf} divided by the partial factors $\gamma_m\gamma_{f3}$.

When longitudinal web stiffeners are provided, the treatment in the code allows a deep web to contribute either fully or partially to the effective section in bending, depending on the amount of stiffening provided. Clause 3/9.4.2.5.2 allows a fully effective web thickness to be used in the determination of the effective section for bending stress analysis (Clause 3/9.4.2). However, it may not be economic to provide sufficient longitudinal stiffening for each of the web panels to resist the stresses associated with this fully effective web. The designer is therefore permitted to shed some of the load in any web panel and to redistribute this load into the remainder of the section (Clause 3/9.5.4). In effect this allows a reduced effective thickness (t_{we}) for the web panel in determining the effective section in bending (but see further comment in Section 5.5.2).

5.5.2 Resistance of web panels

Each web panel (bounded by transverse and longitudinal stiffeners) has to be sufficiently strong to carry the direct and shear stresses on it, as determined from the section analysis and allowing for any redistribution. The checks for yield consider the equivalent stress near, but not at, the boundary where the longitudinal stress is higher (Clause 3/9.11.3). The checks for buckling consider direct stresses in both directions in combination with shear stresses (Clause 9.11.4).

Note that the allowance for reduced effective thickness of the web (Clause 3/9.5.4) may only apply to the check against yielding, because in a web panel that is unrestrained (which might be the case for a panel adjacent to a thin flange) the reduction is removed by the $(1-\rho)$ parameter in Clause 9.11.4.4 (which checks buckling).

The code does not invoke the post-buckling tension field action that is allowed in unstiffened beams. This has been shown experimentally to be conservative.

5.5.3 Resistance of web stiffeners

The strength of a web stiffener (transverse or longitudinal) and its associated width of web plate must be adequate to resist buckling. Transverse stiffeners are checked over their full height; longitudinal stiffeners are checked over their length between transverse stiffeners. See further guidance in Section 6.4.

5.6 Load-carrying transverse beams

Transverse beams that span between main beams and act compositely with the deck slab are usually designed as non-compact beams in accordance with Clause 3/9.9. In addition to direct traffic loading, transverse beams must be designed for forces associated with their function as transverse bracing.

Transverse beams that act as a crosshead to spread the reaction from one bearing to two main beams are particularly heavily loaded. Such beams may also be connected to the deck slab. The detail of the site connections of such beams needs careful consideration, and the cruciform details that result at the junctions with the longitudinal beams require very careful attention to material properties and welding details. Crosshead beams are usually designed as non-compact sections in accordance with Clause 3/9.9.

For crosshead beams, the torsional restraint required at supports by Clause 3/9.12.5.1 is not provided at the support itself but by the torsional restraint where the crosshead is attached to the main beams. The restraint system should be checked for the required stiffness and strength.

5.7 Fatigue considerations

The fatigue endurance of the steelwork in a bridge is assessed using Part 10 of BS 5400. Fatigue failure of steel depends on the propagation of cracks in regions that are subject to fluctuating stress. The fatigue life depends on the size of the initial imperfection or stress concentration and on the range of the stress variation.

Part 10 allows for the size of imperfection or stress concentration by means of a comprehensive classification of welded and non-welded details (Table 17). The designer simply identifies the appropriate classification for the detail being considered.

5.7.1 Methods of assessment

The Code provides three methods for the assessment of fatigue life. The methods involve different determinations of the effective range of stress variation. In order of increasing complexity, they are:

- Without damage calculation – limiting stress range (Clause 8.2).
- Damage calculation – single vehicle method (Clause 8.3).
- Damage calculation – vehicle spectrum method (Clause 8.4).

The first method is most commonly used and is much the quickest, though somewhat conservative. In some cases, where this method indicates failure by a small margin, designers may opt to recheck by the second method. The third method would normally be used on large and complex structures where economy requires greater precision of assessment.

The method given in Clause 10/8.2 requires the determination of the maximum and minimum stresses as a 'standard fatigue vehicle' (Clause 10/7.2.2.1) crosses the bridge. The range (maximum to minimum) is then compared with a limiting stress range appropriate to the classification detail and the spectrum for the road category (Clause 10/8.2.2 and 10/Figure 8). If designers find that the range exceeds the limiting range, they may choose to specify a better class of detail, to modify the design to reduce the stress range or to re-assess by the Clause 8.3 method.

The method given in Clause 10/8.3 requires the determination of a number of different ranges, of differing magnitude, as the standard fatigue vehicle transverses the length of the bridge. Fatigue life is then assessed by summing the damage caused by repeated application of all these ranges. It therefore involves more calculation but is less conservative.

The method given in Clause 10/8.4 is a still more detailed method that requires calculations for a spectrum of different vehicles.

5.7.2 Classification of details

Fatigue detail classifications relate to the potential imperfections at welds, holes or other discontinuities, and their relationship to the direction of the stress variation. The greater the imperfection the lower the stress range that can be tolerated for a given fatigue life. The complete range of classifications is shown in Table 17 of Part 10. This Table shows a range of typical details in classes A to G, plus W (for welds) and S (for shear connectors). For a given stress range, class A has the highest fatigue life, class W the lowest (class S is a special case, see further discussion in Section 7.5).

The attachment of web stiffeners or other elements not carrying load in the stressed direction, produces a Class F fatigue detail. Reinforcing plates and bearing plates welded to the underside of the flange introduce a Class G detail, as do other attachments that are close to the edge of the flange. Shear connectors produce a Class F detail in the stressed direction of the flange plate (as well as the Class S detail that requires special attention). Grip bolted splices introduce Class D details. Cross bracing introduces a variety of details, including Class W, in respect of the transverse action.

To ensure that the workmanship will be appropriate to the fatigue life and detail class assumed by the designer, Part 10 recommends that all areas where class F or better is necessary should be marked on the drawings. The marking should show the fatigue class and the direction of stress to which it relates (see Clause 10/5.3.1). Workmanship levels, which are specified in Part 6, relate the level of inspection and the acceptance criteria for imperfections to the minimum fatigue class required by the designer.

Because of the details that are introduced by attachments, the regions that usually require consideration of fatigue life include: webs and flanges over internal supports; at the attachment of web stiffeners, flange or web reinforcing plates; all connections in transverse bracing and at splices.

Note that Advice Note BA9/81, as amended in November 1983, recommends that the design fatigue life determined according to the rules in BS 5400-10 be reduced for material over 12 mm thick. A reduction factor is given and this can be converted to a limiting stress range. The stress range is reduced typically by about 10% for 50 mm thick flanges.

5.8 Selection of steel subgrade

All parts of structural steelwork are required to have adequate notch toughness, to avoid the possibility of brittle fracture (Clause 3/6.5). Brittle fracture can initiate from a stress concentration when loading is applied suddenly, if the material is not sufficiently 'tough'. The degree of toughness required is expressed as a Charpy impact value (determined from a test carried out on a sample of material) and the requirement depends on the thickness of the material, its minimum temperature in service, the stress level and rate of loading.

The design requirements are expressed in Part 3 in terms of a limiting thickness for the given conditions and specified material. If the thickness of the part does not exceed the limit, there is sufficient toughness.

The most obvious condition that needs to be considered is the lowest temperature that the steel will experience. Minimum bridge temperatures are specified in Part 2 (Clause 2/5.4.3) and are typically -15°C in England and -18°C in Scotland, for composite bridges.

Other conditions relate to the stresses locally at the detail. Special 'k-factors' to account for stress level, rate of loading, stress concentration and potential facture initiation site are given in Clause 3/6.5.3. Multiplying all four factors together gives a single overall factor k.

Steel material, to one of the Standards listed in Clause 3/6.1.2, will have a specified minimum Charpy impact value at a given test temperature. Commonly, steel grade S355J2 to BS EN 10025 is chosen for bridge steelwork, and this has an impact energy of 27 J at −20°C. (The impact energy is indicated by the 'J2' part of the designation.)

The 'maximum permitted thickness' of the steel part is given in Clause 3/6.5.4 in terms of an equation relating the overall factor k and design minimum temperature, the chosen material yield strength and the Charpy test temperature for the chosen material grade.

For the specific case of $k = 1.0$, a simple tabular presentation of limiting thickness for the grades of material covered by Clause 3/6.1.2 is also given.

If the actual thickness of the part exceeds the limiting thickness for the material grade, either a tougher grade must be selected or the detail must be revised. As mentioned above, J2 grade is commonly used for steel to BS EN 10025, but thick parts may need to be grade K2 while thin parts could be J0. It is not necessary to use the same grade throughout the whole structure – thick tension flanges could be K2 grade while the remainder is J2, for example – but the designer should be aware of the possibilities for confusion if more than one grade is used.

6 DETAILED DESIGN: BRACING, STIFFENERS AND SPLICES

6.1 Bracing

As explained in Section 5.3.7, bracing is required at the supports to restrain the primary members (the longitudinal beams) and to provide a load path for transferring horizontal forces (transverse to the main beams) to the level of the bearings, where restraint is provided by the substructure.

Bracing is usually also provided at intermediate positions, for two purposes: to restrain the compression flange against buckling and to act as a load path for horizontal forces transverse to the longitudinal beams. Restraint is likely to be needed to top flanges in midspan regions during construction and to bottom flanges adjacent to intermediate supports.

6.2 Bracing: restraints at supports

Restraints at supports are provided either by triangulated bracing systems or by stiff beams. The bracing systems at the end supports are usually also required to support the end of the deck slab, even when there are no deck cross-beams elsewhere. Typical arrangements for bracing when the span is square to the supports, and there are bearings under each beam, are shown in Figure 6.1.

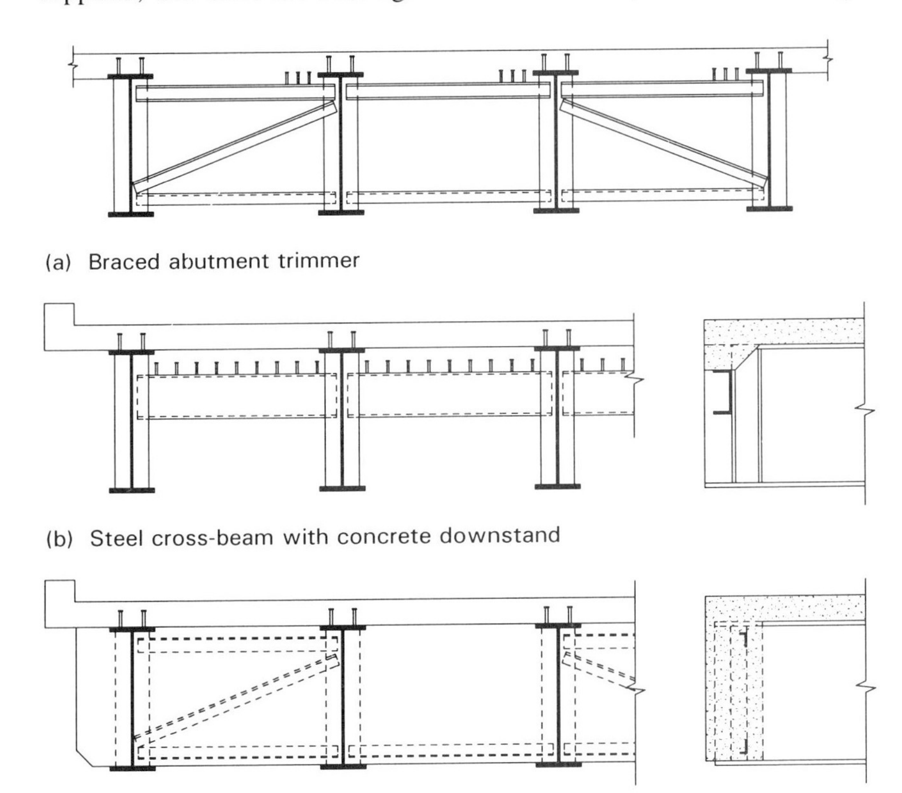

(a) Braced abutment trimmer

(b) Steel cross-beam with concrete downstand

(c) Concrete diaphragm (with bracing for construction condition)

Figure 6.1 *Commonly used bracing systems at supports*

Bracing members are usually angle sections, and the bracing system is continuous across all the beams. Usually only one bearing is restrained laterally, so the continuity of the bracing system ensures that the transfer of horizontal load is shared by the whole bracing system.

At the end supports (abutments) the trimmer beam that supports the end of the deck slab is often encased in concrete. Sometimes the whole bracing system at the abutment is provided by a reinforced concrete end beam, although this can complicate construction because there is no restraint until the beam has been cast.

When the bridge is skew, the question of how to align the support bracing system needs careful consideration. For modest skews (up to about 20°), the bracing may be skewed so that it is in line with all the support bearings. In designing the bracing, account should then be taken of the slight modification to effective torsional stiffness of the restraint to the main beams. Account must also be taken of the twists of the main beams (about their longitudinal axes) that arise as a result of rotation of the bracing planes as the beams deflect (particularly during construction - see Guidance Note 7.03).

When the skew is large, it may be better to provide bracing square to the beams, which means that bracing between a pair of beams connects at a bearing position on one beam and at a position a little way into the span on the other beam. Again, the effects of twist of the main beams (this time because the end which is in the span can deflect vertically) must be taken into account. If square bracing systems are provided at end supports, a tie member may still be needed along the line of bearings, to ensure sharing of lateral loads.

The design requirements for restraint at supports are given in Clause 3/9.12.5. This defines the couple that is needed to provide adequate torsional restraint to the main beam and which must be resisted by the support bracing system at the same time as other applied forces. The required restraint force depends on numerous factors, including the slenderness of the beam and the verticality of the web.

The members forming the bracing system should be designed to the appropriate Sections of Part 3 (3/9, 3/10, 3/11, 3/14).

Restraint to the main beams is also provided by 'crossheads' when these are used to support the bridge between main beams, rather than directly under main beams. Crossheads are substantial primary members that must be designed for moment and shear in a similar manner to longitudinal beams. Because they are substantial, they normally provide the necessary restraint to the ends of the longitudinal beams without specifically being designed for that purpose. The design of crossheads is outside the scope of this publication; some general guidance will be given in Guidance Note 2.09 (not yet published).

6.3 Bracing within the span

As explained in Section 5.4.2, restraint to the main beams within the span, to limit slenderness, is considered by Part 3 to be provided in a number of different ways. One of these is direct connection of the slab to the compression flange and one other (continuous U-frame restraint) does not involve bracing. There are three bracing systems that provide the other forms of restraint:

- discrete lateral restraints
- discrete torsional restraints
- discrete U-frame restraints.

Design aspects for bracing within the span are discussed in Sections 6.3.2 to 6.3.4 below. In all cases, the members forming the bracing system should be designed to the appropriate Sections of Part 3 (3/9, 3/10, 3/11, 3/14).

6.3.1 Forms of bracing

Lateral restraints

Discrete lateral restraint is provided by plan bracing to the compression flange, or by transverse triangulated frames in conjunction with either plan bracing or a deck slab attached to the tension flange.

Plan bracing is usually only provided when there is no deck slab or other means to carry lateral loads back to the supports (lateral bending resistance of the main beams is sufficient in many cases, especially when the load can be shared between many main beams). Before the slab is cast, plan bracing to the top flange will provide the necessary restraint for single spans but, for continuous spans, transverse bracing will also be needed adjacent to the intermediate supports to restrain the bottom flange. Plan bracing to the bottom flange may also be used in longer spans where the designer chooses to provide lateral restraint to the compression flange (adjacent to the supports) at discrete points between effective transverse bracing (rather than increase the number of transverse bracings). For spans over about 60 m, plan bracing over the full length of the bottom flange may need to be provided as a means to increase torsional stiffness (effectively creating a box section) and thus improve dynamic performance (to resist wind-induced oscillations).

In most cases, direct connection of plan bracing to the flange plate should be avoided; instead, it should be connected to horizontal cleats between web and transverse stiffener, just clear of the flange, or to cleats on top of the top flange but within the slab between the layers of reinforcement. Plan bracing below the top flange should be sufficiently clear to allow for the slab formwork. Plan bracing within the slab is preferable for durability reasons but it complicates the fixing of reinforcement and is therefore disliked by some contractors.

Typical triangulated transverse bracing systems are shown in Figure 6.2.

A triangulated 'X' system is suitable for deep main beams and a 'K' system is useful when an X would be very flat (the K system will be stiffer). In a triangulated system, the position of any horizontal members at the top flange should be considered carefully because they may conflict with supports to the formwork for the deck slab. These members are usually redundant once the slab has been cast but removal involves an additional operation (with potential hazards for the operatives), so they are usually left in place. Bracing members are usually lapped and bolted to web stiffeners; welding is rarely used.

As mentioned in Section 5.3.7, intermediate bracing that is continuous across more than two main beams will participate in the global action and will distribute load to several main beams. However, such continuity does not provide much benefit to the design of the main beams but introduces stress reversals in the bracing and

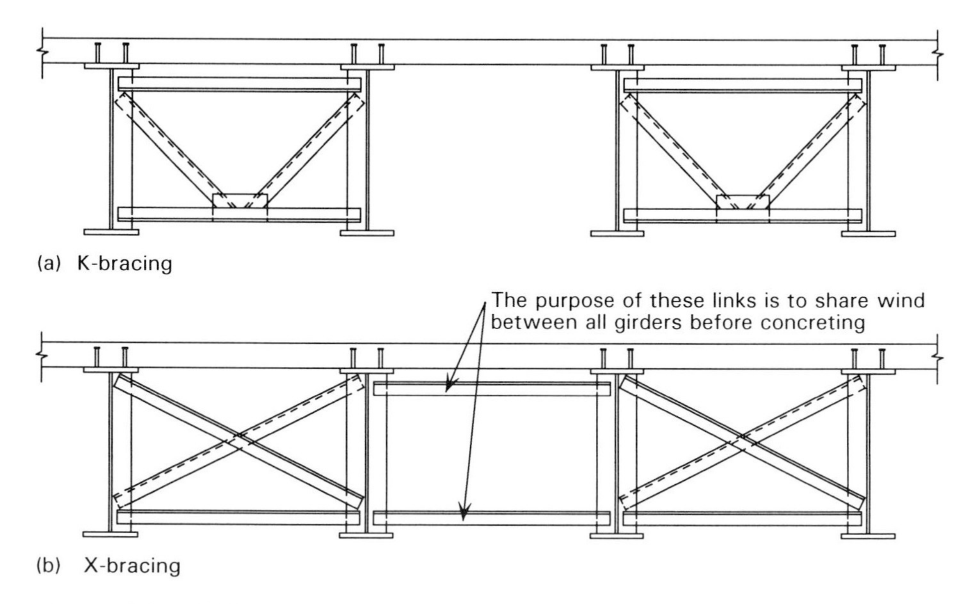

Figure 6.2 *Triangulated transverse bracing systems*

its connections; checks must therefore be made for fatigue. To avoid this fatigue situation, designers frequently use non-continuous bracing, where main beams are connected in pairs, with no bracing between one pair and the next. When continuity is required only during construction (for example to share wind loads between all the beams), temporary bracing is sometimes provided between adjacent pairs of beams (see Figure 6.2).

Torsional restraints

During construction, when there is no deck slab, and where there is no plan bracing to provide lateral restraints, torsional restraint to the beams can be provided by triangulated frames (similar to those shown in Figure 6.2) between main beams or by stiff beams that are rigidly connected to stiffeners on the webs of the main beams.

Torsional bracing using stiff beams, in the form of a channel or similar deep member (thus often called 'channel bracing'), is suitable for relatively shallow main beams (up to about 1200 mm deep). The beam is usually lapped and bolted to the web stiffeners; often the ends of the beams have deep gusset plates welded to them so that the bolt group has a higher moment capacity. See Figure 6.3.

Once the deck slab has been cast, torsional bracing effectively becomes part of a discrete lateral bracing system.

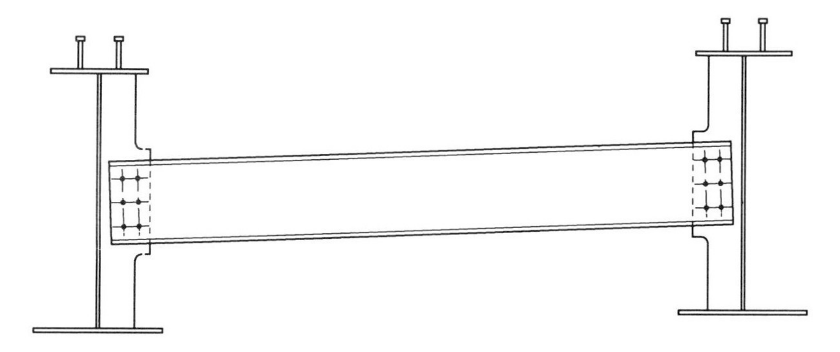

Figure 6.3 *Torsional restraint by channel bracing*

Discrete U-frame restraints

Discrete U-frames are created by stiff transverse beams, connected close to the tension flange, acting in conjunction with either plan bracing or deck slab at tension flange level, and with stiffeners on the webs of the beams.

When U-frames created by ladder-deck construction provide restraint to the bottom compression flange, a moment connection is needed between cross-beam and stiffener. When the main beams are very deep, 'knee bracing' is sometimes provided to give a better and stiffer restraint to the compression flange.

6.3.2 Design of discrete lateral restraints

Elements providing discrete intermediate restraints to compression flanges should be checked at ULS in accordance with Clause 3/9.12.2.

The cases where discrete lateral bracing is provided include:

- where plan bracing is directly connected to the compression flange (midspan regions during construction)
- where plan bracing is connected to the tension flange and stiff (triangulated) transverse frames are provided at the positions where the plan bracing is connected (support regions during construction)
- where the deck slab is connected to the tension flange and stiff transverse frames are provided between compression and tension flanges (support regions of continuous bridges after the slab has been cast).

In these cases, the plan bracing members and any transverse bracing frames should be designed to resist a lateral shear force F_R given by Clause 3/9.12.2, in conjunction with wind or other forces. Note, though, that the force F_R is an internal self equilibrating force and should not, therefore, be included in the design loads on the bearings.

The magnitude of the force F_R depends on the stiffness of the restraint provided (subject to an upper limit) magnified by a factor to allow for the proportion of elastic critical buckling load (in lateral torsional buckling) that is being carried.

6.3.3 Design of discrete torsional restraints

Torsional bracing between a pair of beams, without any plan bracing or deck slab (i.e. during construction) does provide restraint against LTB (see Section 5.4.2).

In such cases, each restraint should be sufficiently stiff and be capable of resisting a pair of equal and opposite forces F_R (i.e. a couple), as required by Clause 3/9.12.2. Because these restraints do not provide any lateral restraint, the beams connected by such bracing must be designed to resist lateral forces (such as wind and construction sway loads) by bending in plan at the same time as they carry the vertical bending effects.

6.3.4 Design of discrete intermediate U-frame restraints

Discrete U-frames in conjunction with some form of plan bracing at the level of the cross members provide flexible restraint to a compression flange. [U-frames without plan bracing would have to be considered as torsional restraints without bracing, in accordance with Clause 3/9.12.2(b)].

For discrete U-frames, the design forces F_R for the U-frames are given by Clause 3/9.12.3.2, in the same manner as Clause 3/9.12.2 for intermediate restraints, except that the stiffness used in the calculation is that of the U-frame, as defined in Clause 3/9.6.4.1.3.

Where the U-frame cross member is subject to vertical loading (for beam on slab construction this means U-frames adjacent to supports) an additional force F_C must be calculated (Clause 3/9.12.3.3).

During construction, before the slab is cast, transverse beams moment-connected to the main beams are only considered as discrete U-frames when there is plan bracing.

6.3.5 Design of continuous U-frame restraint

Where there is no discrete restraint to the bottom flange adjacent to supports, continuous restraint is effectively provided by the U-frame action of the deck and the girder web. In that case, the web must be checked for the effects of a continuous lateral force. The magnitudes of the force f_R and a force f_C due to the bending of the deck slab are given by Clause 3/9.12.4.2. These forces are determined in a similar way to those for the discrete U-frames.

6.3.6 Intermediate bracing in skew spans

Restraint to the main beams is needed orthogonally to the beam axis. Where this is provided by bracing between beams, this is best achieved by using planes that are square to the main beams, even when the bridge is skew. This will mean that the effective lengths of the beams adjacent to a support are not all the same but this does not raise any problems in principle.

Simple lateral ties to share wind load during construction (such as those shown in Figure 6.2) could be on the skew, linking two staggered planes of bracing, provided that the skew is small.

6.4 Web stiffeners

Web stiffeners are required to improve the shear resistance of the web, for the attachment of transverse bracing and as support over bearings.

6.4.1 Intermediate transverse web stiffeners

Intermediate web stiffeners are usually made by welding a simple flat plate to one face of the web. The outstand proportions of the flat are limited by Clause 3/9.3.4, to ensure that its strength is not limited by local buckling.

The outstand and a portion of the web plate on either side form an effective Tee section (see Clause 3/9.13.2) that has its centroid just outside the face of the web.

Intermediate web stiffeners should be designed to resist:

- axial force due to tension field action
- forces and moments due to action with transverse bracing system
- direct loading (from a wheel at the stiffener position)

- bending about a longitudinal axis due to eccentricity of axial forces (in the plane of the web) relative to the centroid of the stiffener section
- destabilising influence of the web (buckling check only).

Most of the forces that the effective stiffener section is required to resist arise in the plane of the web, so they cause both axial and bending stresses in the effective section.

To restrain the web, the stiffener does not need to be connected to either flange, although it is usual to connect it to at least the top flange. Intermediate stiffeners are frequently not connected to the bottom flange. Where the stiffener is not connected, the code specifies a maximum clearance of five times the web thickness, but it is suggested that a clearance of not more than three times should be aimed for.

Where the stiffener acts as a connection for transverse bracing, it should be connected to both flanges. Practical experience from North America indicates that failure to provide such attachment may lead to fatigue cracking in the web at the point of curtailment of the stiffener.

Where bracing is attached to a web stiffener, the stiffener may need to be shaped to provide sufficient lap to connect the bracing members.

The fillet welding of the end of a stiffener to a flange does not introduce a lower class of fatigue detail than is likely to be present already, provided that the toe of the weld is at least 10 mm from the edge of the flange. Web stiffeners should be proportioned such that they are narrower than the flange outstand; shaped stiffeners for bracing may need to be notched at the end to ensure that the welds are not too close to the edge of the flange.

Intermediate stiffeners should normally all be attached to one face of each beam web. On the outermost beams, the stiffeners should be on the hidden face, rather than the exposed face, for better appearance.

Where a sloping flange changes direction (at the end of a tapered haunch), a transverse stiffener is required to carry the transverse component of force (Clause 3/9.13.1). Although all transverse stiffeners are strictly required to be full depth, it is common to make such stiffeners part height, provided that they are symmetrical about the web and the load can be satisfactorily transferred into the web.

6.4.2 Load-bearing web stiffeners

Bearing stiffeners are effectively a special case of transverse web stiffener, and the design rules in Clause 3/9.14 contain many similarities to the rules in Clause 3/9.13. Again, flat plate stiffeners are often used, although more than one flat on each side may be needed, partly to provide sufficient area and partly to provide greater resistance to reactions that are eccentric along the beam.

Bearing stiffeners are usually provided on both faces of a web, and are usually symmetrical about the web, so that the centroid of the effective section is on the line of the web. In some cases, it may be convenient to provide two stiffeners on one face and a single stiffener on the opposite face of the web. This would give resistance to bending about a transverse axis while allowing a single lapped connection to transverse bracing. See Figure 6.4 for typical details.

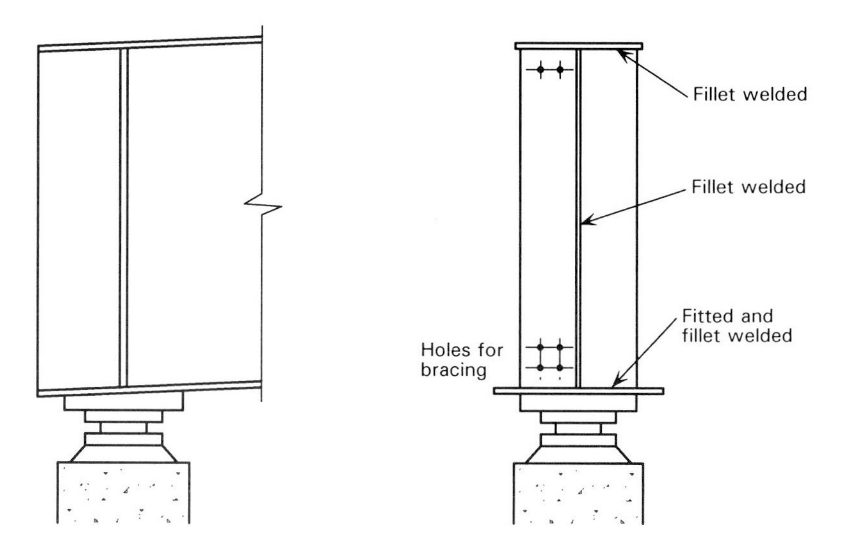

Figure 6.4 *Typical bearing stiffener*

Bearing stiffeners generally need only be checked for the ULS, in accordance with Clause 3/9.14, though the fatigue endurance of certain details must also be checked. The effective stiffener section must be able to resist:

- axial force arising from the support reaction
- forces and moments due to action in conjunction with cross-beams or bracing system (including effects of the couple needed to restrain the beam in torsion)
- bending (about longitudinal and transverse axes) due to eccentricity of bearing reaction relative to the centroid of the stiffener section
- direct loading (from a wheel over the bearing stiffener)
- destabilising influence of the web (this need only be considered for the check against buckling).

Axial load and bending due to bearing eccentricity are assumed to vary linearly to zero at the level of the top flange.

Some values to be used for eccentricity of bearings are given in Clause 3/9.14.3.3 but it must be noted that these are not exhaustive. They are applicable where radiused or flat-topped bearings are installed accurately relative to the beam steelwork. No values are given for elastomeric pot bearings, which are in common use; it is suggested that a value of 10 mm in either direction be used for the unevenness or inaccuracy of such bearings. Movements of the beam relative to the bearing due to changes in temperature are readily calculated. Further consideration should, however, be given to eccentricities due to fabrication tolerance, particularly on long viaducts, if it is not certain that the lower part of the bearing can be positioned accurately relative to the girder after erection, and to eccentricities due to shortening as a result of shrinkage of the concrete.

Three checks for adequacy are required:

- on direct bearing at the bottom of the web and stiffeners
- on maximum stresses on the effective stiffener section
- on buckling of the effective stiffener as a strut between flanges.

Bearing stiffeners should be 'adequately connected to both flanges' and should be 'fitted closely to the flange ... subject to a concentrated load' (Clause 3/9.14.1). Close fitting to the bottom flange allows the compressive stresses to be transmitted in bearing rather than through welds, and should therefore usually be specified (ensure that it is marked on the drawings). Close fitting to the top flange is not necessary and should not be specified (to ensure that it does not compromise fitting of the bottom flange); welding is an adequate connection to the top flange.

Although a fitted connection at the bottom will transmit compression in bearing, it should be noted that for the fatigue check all the variation in reaction due to the fatigue vehicle load is assumed to be transmitted through the welds (see Clause 3/9.14.4.1) in a Class W detail (see Part 10, detail 3.11). As for intermediate stiffeners, bearing stiffeners should be sized to ensure that the weld toes are not within 10 mm of the edge of the flange plate, or a lower class fatigue detail will result.

For further guidance on detailing and fabrication of bearing stiffeners, see Guidance Note 2.04.

It should be noted that when the main beam is designed as a compact section over intermediate supports, plastic redistribution of stresses is assumed and consequently the check on yielding of the web plate (Clause 3/9.14.4.1) is inappropriate. (The combined action permitted by Clause 3/9.9.3.1 allows full longitudinal yield in tension and compression at the same time as up to half shear yield - the equivalent stress is then in excess of yield throughout the depth of the web, before any consideration of the transverse stress associated with the bearing stiffener.) The yield check may therefore be ignored at ULS.

6.4.3 Longitudinal stiffeners

The provision of longitudinal stiffeners on a deep slender web has a significant effect on the form of web buckling that can develop and, as mentioned previously, the code then requires each of the panels bounded by stiffeners to be checked separately. It requires also that the effective web stiffener (stiffener plus a width of plate) is an adequate restraint to the edge of the panel. This requirement is expressed in Clause 3/9.11.5.

Discontinuous longitudinal stiffeners (i.e. stiffeners that stop just short of the transverse stiffeners) are effective in restricting web buckling. They are commonly used to simplify fabrication and reduce costs. This is permitted, though the stiffener must not then be included in the effective section in bending (Clause 3/9.4.2.6), although, for the design of the stiffener itself, it is assumed to be at the same longitudinal stress as the unstiffened web at that position. Further, the stress concentration that is caused by the discontinuity needs to be checked, particularly for fatigue.

Stiffeners that are continuous and assumed to be contributing to the bending resistance should extend beyond the point at which the contribution is needed (Clause 3/9.11.6).

6.5 Connections and splices

6.5.1 Bolting

Bolted connections are usually connected using High Strength Friction Grip (HSFG) bolts, manufactured in accordance with BS 4395[23] and installed in accordance with BS 4604[24]. Black bolts are not permitted in any structural connection (Clause 3/14.5.3.1). The use of higher grade bolts (e.g. Grade 8.8) to BS 3692 in clearance holes is considered in the same category as black bolts and is thus prohibited.

When the connection is designed to act in bearing/shear at ULS, the requirement for no slip at SLS usually governs (the chief exception being for thin material when large bolts are used).

For further advice on the design of bolted connections, see Guidance Note 2.06.

6.5.2 Welding

Welded connections are made using either fillet welds or butt welds. Design of welded connections is covered by Clause 3/14.6. Reference is made in that clause to BS 5400-6 (for workmanship) and to BS 5135[25] (for detailing of welds).

Because weld details are a potential source of local defect, the choice of weld detail has a significant effect on fatigue performance. BS 5400-10 classifies weld details according to their effect on fatigue endurance (see Section 5.7). The inspection and testing of welds is an important aspect of quality control in fabrication; see Guidance Notes 6.01 to 6.03 for further advice.

6.5.3 Splices in main girders

Each main girder is fabricated in a number of pieces and joined together on site, either prior to or during erection. The lengths of the pieces are usually chosen to suit economical fabrication, and splice positions are usually arranged to be away from positions of maximum moment. The splice may then be designed simply to transmit the maximum design load effects at that position (Clause 3/14.3.1). Splices may be bolted or welded.

At a HSFG bolted splice, cover plates are normally provided on both faces of each flange and web. The number of bolts required may be determined in accordance with Clause 3/14.5.4, either on the basis of friction capacity at ULS or, more economically, on the basis of no slip at SLS and bearing/shear at ULS. Note that this second method cannot be used when holes are slotted or oversize; design must then be on the friction capacity at ULS. Bolts should be spaced in accordance with Clause 3/14.5.1. Stresses should be checked in the cover plates and in the girder on the weaker side of the splice, allowing for holes in determining net sections (see Clause 3/14.4).

It may be necessary to provide shear connectors on the upper cover plate to the top flange, to comply with maximum spacing limitations. The number of connectors should be kept to a minimum, as it complicates the tightening of the bolts.

A typical bolted splice is shown in Figure 6.5.

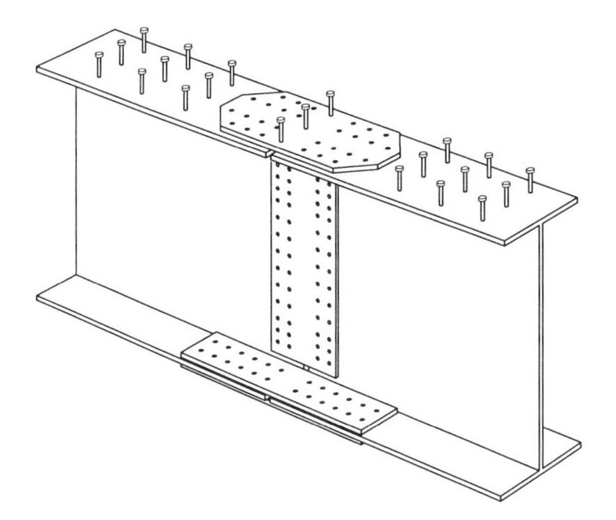

Figure 6.5 *Typical bolted splice*

Welded splices in girders usually involve full penetration butt welds in web and flanges. Partial penetration welds are not permitted to be used to transmit tensile forces (Clause 14.6.2.2).

6.6 Bearings

Bearings are normally selected from proprietary product ranges to suit the particular requirements of the bridge. The designer must therefore determine the range of effects on bearings before the selection can be made.

The effects to be considered in the selection of bearings are:

- vertical and horizontal loads
- vertical and horizontal movements
- transverse and longitudinal rotations.

All the effects, determined from the global analysis, should be summarised on a bearing schedule, such as that given in Table 9 of Part 9.1. Once the effects have been entered, the schedule may be completed by entering details of the selected bearings.

There is a wide range of bearings to choose from:

- laminated elastomeric bearings
- pot bearings
- rocker bearings
- spherical bearings
- tie-down and other special bearings.

In most bridges, the bearings are located below the bottom flange of the girder, supported directly by the abutment or pier.

Moving bearings and bearings with rubber-based components usually require some maintenance during the life of the bridge. The designer should make provision in the design for the girders to be jacked off the bearings so that maintenance can be carried out. Inspection and maintenance of bearings is improved by placing them on plinths above the general level of the abutment seat.

The selection or design of bearings is outside the scope of this publication. Advice can be found in *Bridge bearings and expansion joints*[2], in the *Steel designers manual*[26] and in Guidance Note 3.03. Advice on the attachment of bearings is given in Guidance Note 2.08.

7 DETAILED DESIGN: DECK SLAB AND SHEAR CONNECTION

The comments in Sections 7.1 to 7.3 relate principally to deck slabs on multiple beams. Additional considerations for ladder beam decks are discussed in Section 7.4.

7.1 Load effects in deck slab

The load effects in the slab arise from two sources, the global composite action with the steel girders and the local action in distributing the loads from individual wheels. It is convenient to calculate the effects from the two sources separately.

The load effects in the slab from global action are longitudinal stresses due to bending of the composite beams and transverse stresses due to the bending of the slab as the main beams deflect under load. The moments in the main beams and the transverse bending effects are determined from the grillage analysis (see Section 5.1.1).

To avoid double counting of local effects due to wheel loads, etc., they must be excluded from the global analysis. This is achieved by applying all the loads in the global analysis to the longitudinal members and none to the transverse members. Some grillage analysis programs will automatically distribute patch loads and point loads to the principal longitudinal members, wherever they are applied.

Local effects are principally transverse bending moments that arise from individual wheels. The usual HA loading was derived to produce longitudinal bending moments and shears that are equivalent to the effects of actual axle, bogie and vehicle loads; it is not appropriate for determining transverse effects. To overcome this, transversely spanning slabs are designed for the wheel loads from an HB vehicle (Clause 3/6.4.1.3).

Local effects are calculated by a separate slab analysis, as explained below.

7.2 Local slab analysis

Local analysis for dead load effects can readily be calculated manually; the effects are usually relatively small.

Local moments in the deck slab due to the effects of HA or HB wheel loads can be calculated using recognised methods such as the Westergaard Method[27] or the use of Pucher Influence Charts[28]. These moments are calculated for the slab on rigid vertical supports, making appropriate allowance for continuity of the slab from one slab bay to another with no torsional restraint from the steel girder webs.

The use of Pucher Charts is particularly common. These charts are a series of contour plots of influence surfaces, as illustrated in Figure 7.1. The simplification of support conditions to permit use of standard charts normally leads to a conservative assessment of worst moments.

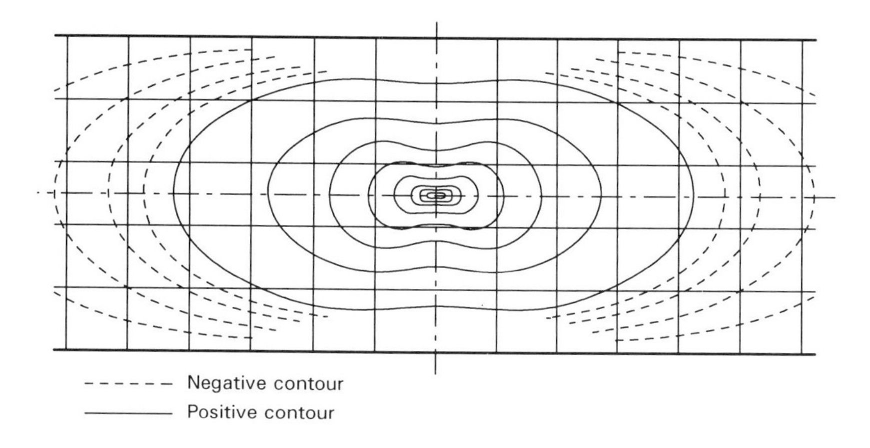

Figure 7.1 *Typical influence chart for slab moments*

As an alternative to separate modelling of local effects, the grillage model can be given a graded mesh, such as shown in Figure 7.2. Wheel loading over the fine mesh will lead to direct values for total moment and twist in the slab. A resolution (node spacing) of about 500 mm should be adequate in view of the loaded area under a wheel and dispersal through surfacing and slab. Note that this alternative is more suitable when the grillage model includes slab elements; if only beam elements are available, a correction for Poisson's ratio effects should be made.

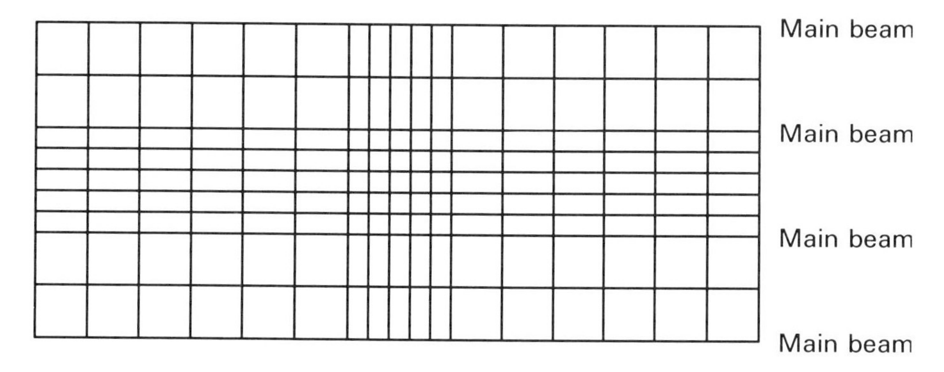

Figure 7.2 *Graded mesh of slab elements*

Such local analysis determines orthogonal moments in the slab. However, where the bridge has a small skew (less than 20°), it is common to place the slab reinforcement parallel to the abutments. In such cases it may be taken that the consequent skewed direction of principal stiffness does not affect significantly the orthogonal local moments (calculated by the above methods).

When the grillage is skewed, distribution moments (which are generally smaller than the local moments) should be resolved into orthogonal directions before adding to local moments.

To design a slab with skewed transverse reinforcement, the moments and twists must be converted into effective design moments in the principal directions. The equations of Wood[29], as modified by Armer[30], may be used for the design of the slab; these are available within some computer analysis programs.

7.3 Slab design considerations

The reinforced concrete deck slab participates in the bending resistance of the main beams and also acts locally in bending to distribute the wheel loads. The requirements for limiting stresses and crack widths must therefore consider the combination of the two effects, where required.

Compressive stresses in concrete and tensile stresses in reinforcement need to be checked at both limit states. At ULS, local bending and global bending may be checked separately (Clause 5/6.1.2).

At the SLS, stresses from the two actions must be added together (Clause 5/5.2.4.1); total stresses and crack widths must comply with relevant limits. (As noted in Section 2.7, all references to Part 5 are to BD 16/82 and the combined document, unless noted otherwise.)

Part 5 requires that global bending stresses be calculated on the effective section, allowing for shear lag at SLS but excluding the effects of shear lag at ULS; this is consistent with the treatment of shear lag in Part 3.

Designers should note the additional requirements in the DMRB for control of early thermal cracking (BD 28/87). This requires, for example, additional reinforcement when a cantilever slab is cast against an existing deck slab.

7.3.1 Limiting stresses in concrete and reinforcement

The limiting stresses (design strengths) for the slab acting compositely in a non-compact section are $0.5\,f_{cu}$ and $0.87\,f_{ry}$ at ULS (Clause 5/6.2.3) and $0.5\,f_{cu}$ and $0.75\,f_{ry}$ at SLS (Clause 5/5.2.2).

The strength of the slab for local effects at ULS and for co-existing effects at SLS is referred to Part 4, which gives the same values of $0.5\,f_{cu}$ and $0.87\,f_{ry}$ at ULS (Clauses 4/4.3.3.3 and 4/4.3.2) and $0.5\,f_{cu}$ and $0.75\,f_{ry}$ at SLS (Clauses 4/4.3.3.2, 4/4.3.1 and 4/Table 2).

For bending resistance of compact sections, the moment capacity is based on a concrete strength of $0.4\,f_{cu}$ or a reinforcement strength of $0.87\,f_{ry}$ (Clauses 4/6.2.2 and 3/9.9.1.2).

In the above discussion, mention of γ_{f3} has been omitted for clarity. It must nevertheless be included in the check for adequacy, where appropriate (see Section 2.2).

7.3.2 Limiting crack widths

Crack width limitations at SLS are given in Clause 5/5.2.6, which refers to Part 4 for calculation of the crack width. Note, however, that where HB loading is to be taken into account, 30 units (not 25) should be used; this change has been made in BD 37/88 (3/6.4.3) and in BD 24/92 (amendment to 4/4.2.2), although BD 16/82 and the combined document (5/5.2.6.2) have not yet been updated.

7.3.3 Longitudinal reinforcement

The minimum longitudinal reinforcement required in each face to satisfy the SLS requirement for control of cracking due to shrinkage and temperature is given in Clause 4/5.8.9 and BD 28/87.

The tensile zone over the intermediate supports may well require an increased amount of reinforcement. The critical crack widths in the slab result from a combination of global strain derived from the composite section analysis and local flexure strain resulting from local longitudinal moments in the deck slab due to wheel loads. This is covered in Clause 4/5.8.8.2(c). If the deck is supported on a cross-beam at intermediate supports, the local longitudinal moments will be hogging; it is clear that the top layer of reinforcement will determine critical crack widths. If there is no cross girder, both hogging and sagging longitudinal moments should be determined and crack widths should be calculated in both the top and bottom faces of the slab. Local longitudinal hogging moments can also occur in decks with torsionally stiff diaphragms or end diaphragms or skew decks.

It should be remembered that local longitudinal moments and strains in the slab are greatest mid-way between the steel girders. However, this is where, because of shear lag, global strains are at their least. The determination of these co-existent stresses due to longitudinal bending is given by Clause 5/5.2.4.3, which refers to Clause 3/A.6.

7.3.4 Transverse reinforcement

The minimum transverse reinforcement required in each face is given in Clause 4/5.8.8.2. Transverse moments over the main beams and midway between the beams will usually require more substantial reinforcement. Transverse reinforcement is usually placed as the outer layer (outside the longitudinal reinforcement) because the transverse moments are greater than the longitudinal moments.

Sagging between the girders will determine the bottom transverse reinforcement; hogging over the beams will determine the top reinforcement. Reinforcement for in-situ slabs is usually uniform across the slab. When permanent precast planks are used, these usually contain additional bottom reinforcement that may be considered to contribute to the transverse sagging bending resistance of the completed deck. By the nature of the construction, these bars cannot be continuous across the beams. Additional transverse bottom reinforcement is required to provide such continuity.

Longitudinal shear between the girder and the slab requires sufficient transverse reinforcement in the bottom face (within the height of the connector) and this may determine the reinforcement near the ends of spans (see Section 5.9).

7.4 Ladder decks

Where ladder deck construction is used, the deck slab responds differently and needs to be designed differently. Each 'panel' of slab is supported on four sides and will therefore be subject to greater biaxial effects than slabs that span transversely between longitudinal beams. Nevertheless, the local loading to be considered for slabs supported on four sides is the same wheel loading as for transversely spanning slabs (see Section 7.1). However, the slab panels in ladder decks generally have a fairly high aspect ratio (typically 9 $\times$ 3 m) and it is arguable that the central portion of the panel (about 6 m for the panel size just quoted) spans principally in the longitudinal direction. In that case HA loading should be considered, and this can be more severe than the HB vehicle for sagging moment at the middle of the panel.

Note that, although the principal local and global effects are both longitudinal for ladder decks, they may be considered separately, as allowed by Clause 5/6.1.2. The loading conditions for maximum effects are quite different for the two cases.

Because the greatest bending moments in the slab are longitudinal, rather than transverse, the longitudinal reinforcement should be placed as the outer layer for ladder decks, except where the slab cantilevers outside the outermost beams without support from cantilever beams.

7.5 Shear connection

Shear connectors are required on the top flange, to provide the necessary shear transfer between the steel girder and the concrete slab that is required for composite action. The shear flow varies along the length of the beam, being highest near the supports, and it is customary to vary the number and spacing of connectors to provide just sufficient shear resistance for economy. The most commonly used form of connector is the headed stud, though channel connectors are sometimes used.

Shear connectors must be designed to provide static strength and for fatigue loading. With non-compact sections, the required resistance at SLS generally governs the design for static strength. Shear flows should be calculated at supports, at midspan and at least one position in between, i.e. quarter points. The ULS need only be considered for non-compact sections when there is uplift or redistribution of tension flange stresses (Clause 5/6.3.4). Fatigue may well govern the spacing of connectors in midspan regions.

For compact sections, because the bending resistance at ULS is calculated in terms of a plastic stress distribution, shear flow at an intermediate support should strictly be compatible with the variation of stresses from the section that is plastic to regions where it is elastic, but that would be complex to evaluate. Instead the code simply requires that the shear resistance be checked at ULS, and by calculating the shear flow on the basis of elastic section properties, and assuming the concrete to be uncracked and unreinforced in both hogging and sagging regions (Clauses 5/6.3.1 and 5/5.3.1). This requirement ensures in practice that there is sufficient connection to develop the plastic moment resistance.

The nominal static strength of shear connectors is given in Part 5, Table 7 and the design static strength in Clauses 5/5.3.3.6 and 5/6.3.4. The design of the connectors must provide a resistance per unit length of at least the maximum design load shear flow over 10% of the length of the span each side of a support. In other parts of the span a series of groups of connectors at constant spacing may be used to provide a 'stepped' resistance, subject to the provision of sufficient total resistance over each length. The maximum calculated shear flow within the length of any such group must not be more than 10% in excess of its design resistance per unit length. In the DTp combined document, the design static strengths are given in modified Clauses 5/5.3.2.5 and 5/6.3.4, and the design procedures in modified Clauses 5/5.3.3.5 and 5/6.3.4.

It should be noted that although the code requires consideration of uplift when there is tension field action (Clause 5/5.3.3.4 or DTp 5/5.3.3.6), there are no provisions for the determination of its value, its point of application or the length over which it may be resisted. Such uplift is usually ignored by designers.

Fatigue design of the shear connectors always uses the simplified procedure of Clause 10/8.2 and checks the stress range in the Class S weld detail between connector and flange (Clause 10/6.4.2), as the standard fatigue vehicle passes along the bridge.

Transverse reinforcement is required in the slab to provide shear resistance at ULS in a similar manner to the requirements for shear stud spacing (Clause 5/6.3.3). The required area of reinforcement is usually provided in multiple beam construction by continuity from midspan of the slab of the bottom layer of transverse reinforcement. In ladder deck construction, the transverse reinforcement required for slab bending is usually smaller and may not be sufficient on its own to transfer high shear flows.

8 FLOW DIAGRAMS

The requirements of BS 5400 for the design of continuous composite beams are presented on the following pages in the form of a series of flow diagrams. There are twelve separate diagrams:

Figure 8.1	An overall diagram, leading to a series of checks for the various structural elements
Figure 8.2	Compact and non-compact beams without longitudinal stiffeners at ULS
Figure 8.3	Beams with longitudinal stiffeners at ULS
Figure 8.4	Compact beams and non-compact beams with high shear lag at SLS
Figure 8.5	Bolted splices at ULS
Figure 8.6	Bolted splices at SLS
Figure 8.7	Longitudinal shear connection
Figure 8.8	Restraints to compression flanges
Figure 8.9	Restraints at supports
Figure 8.10	Bearing stiffeners
Figure 8.11	Intermediate transverse web stiffeners
Figure 8.12	Deck slab

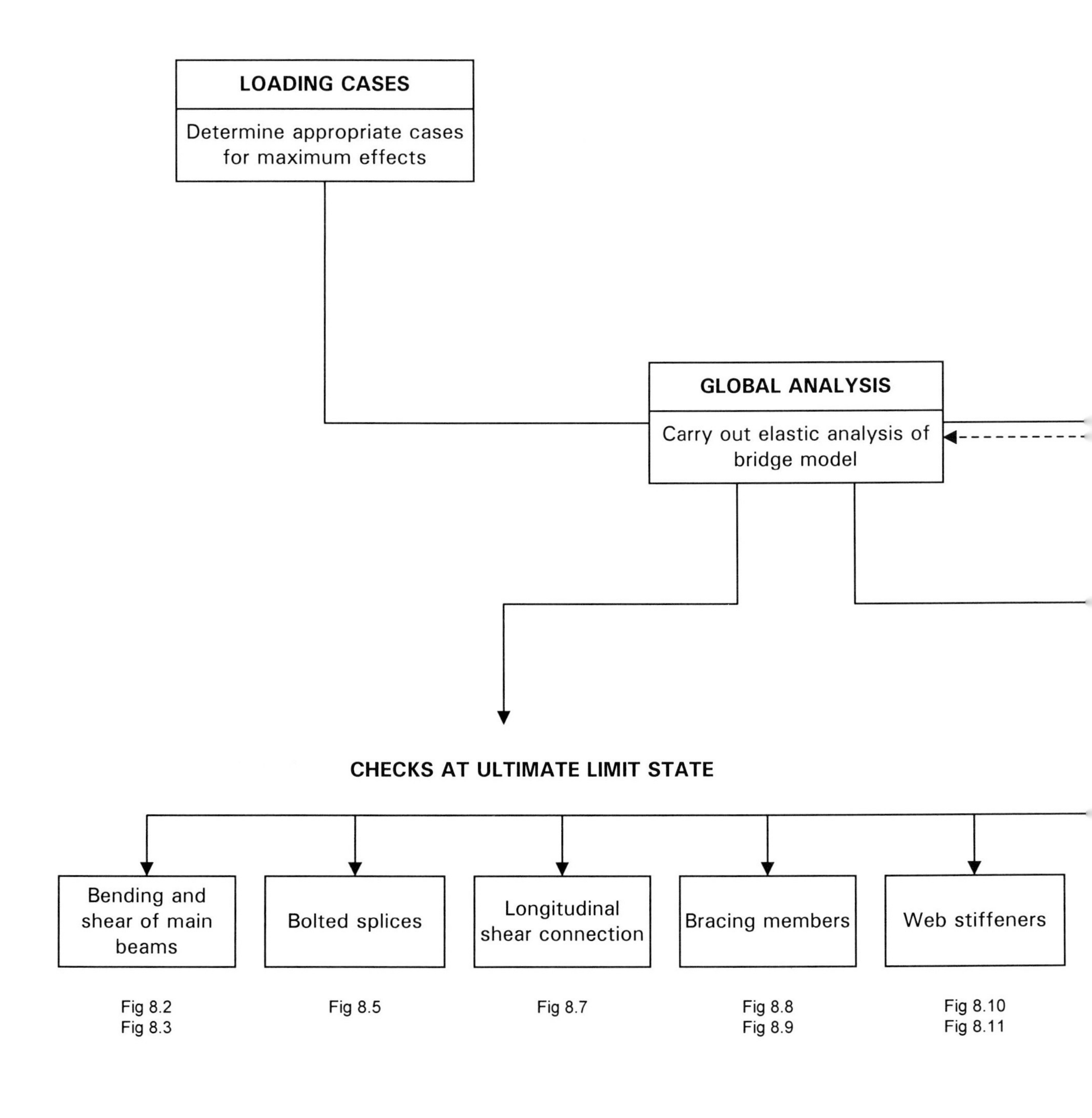

Figure 8.1 *Overall flow diagram for design of composite bridges*

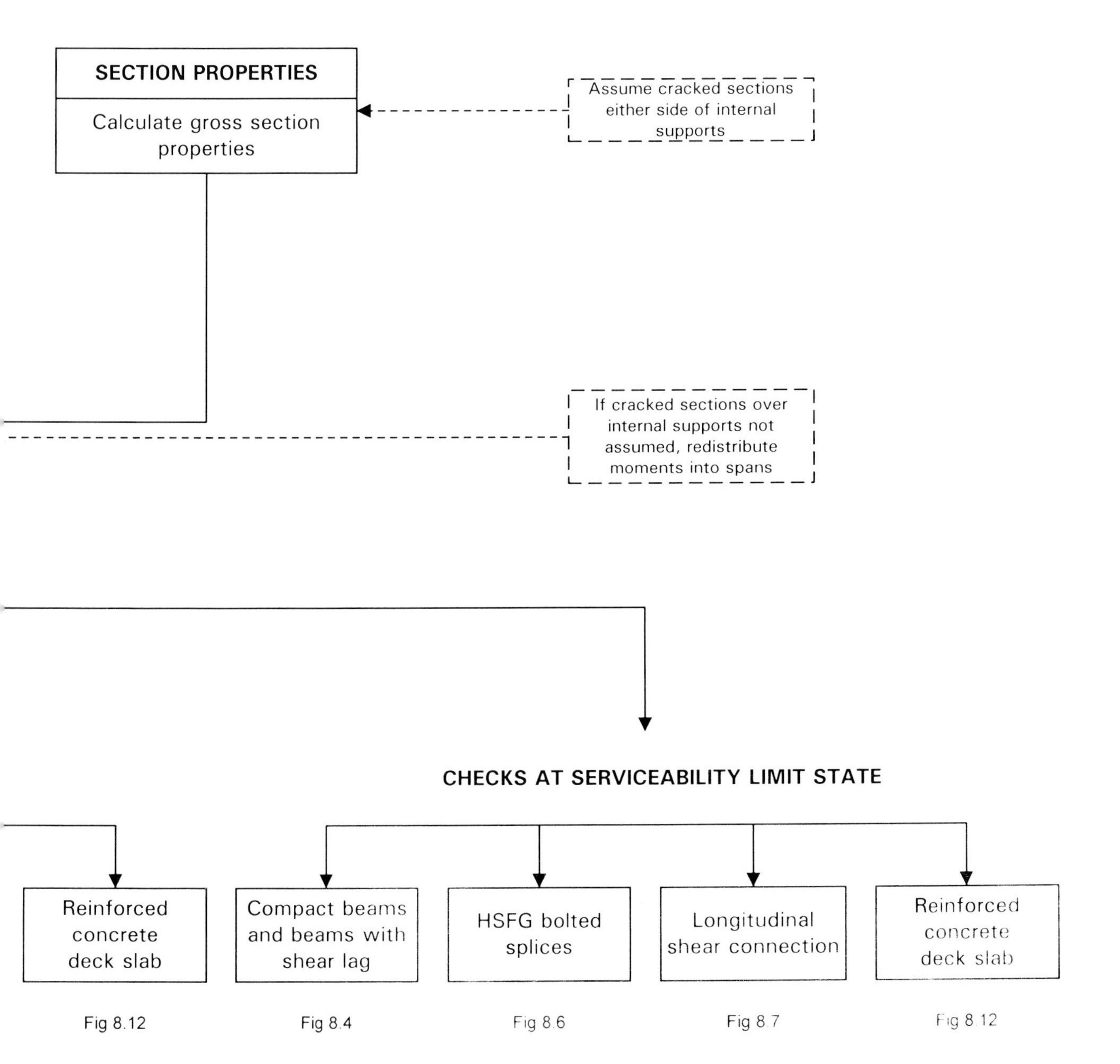
SECTION PROPERTIES
Calculate gross section properties
Assume cracked sections either side of internal supports
If cracked sections over internal supports not assumed, redistribute moments into spans
CHECKS AT SERVICEABILITY LIMIT STATE
Reinforced concrete deck slab
Compact beams and beams with shear lag
HSFG bolted splices
Longitudinal shear connection
Reinforced concrete deck slab
Fig 8.12
Fig 8.4
Fig 8.6
Fig 8.7
Fig 8.12

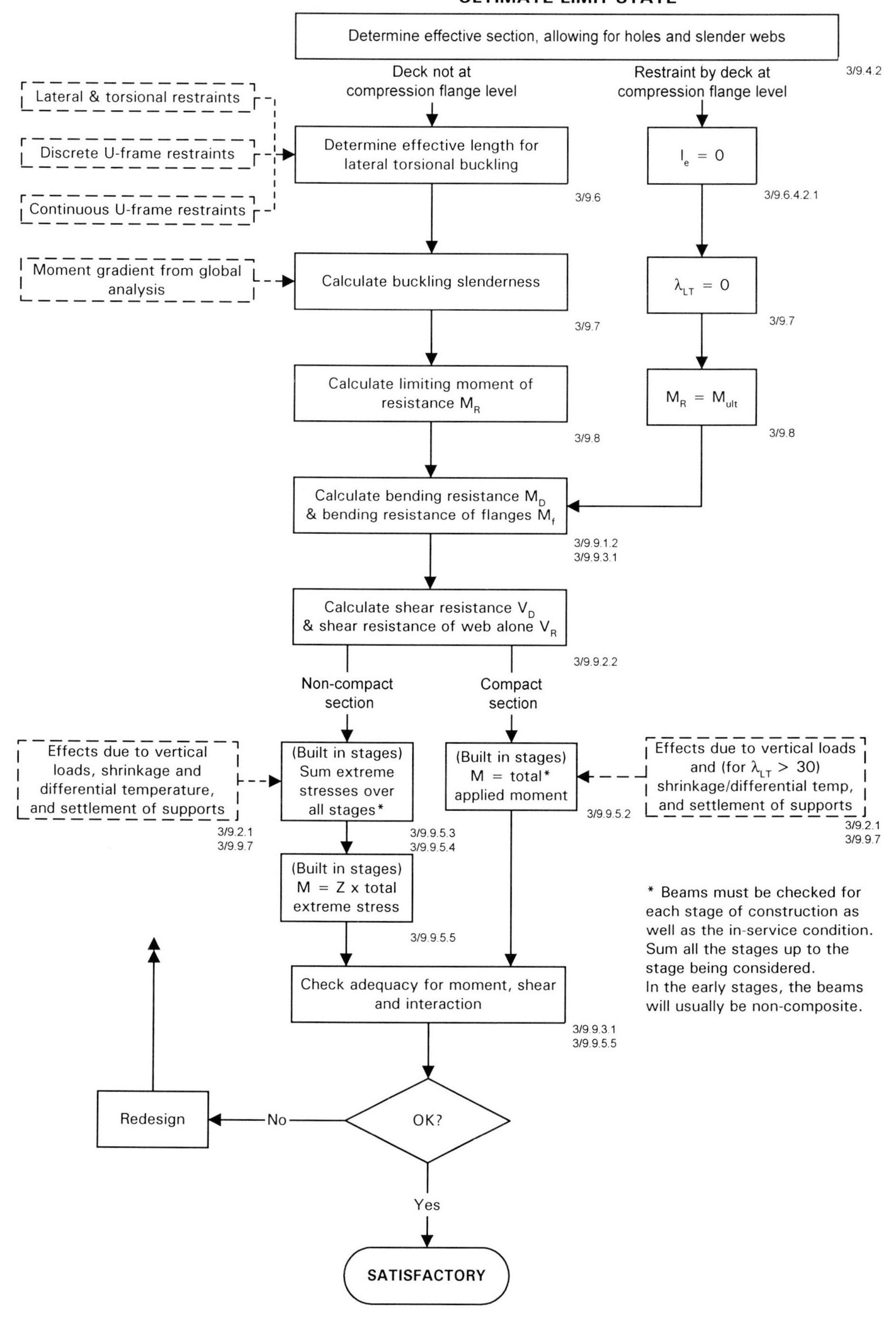

Figure 8.2 *Compact and non-compact beams without longitudinal stiffeners at ULS*

(From Figure 8.1)

ULTIMATE LIMIT STATE

Determine effective section, allowing for holes and slender webs

Deck not at compression flange level

Restraint by deck at compression flange level 3/9.4.2

Lateral & torsional restraints

Discrete U-frame restraints

Continuous U-frame restraints

Determine effective length for lateral torsional buckling 3/9.6

$l_e = 0$ 3/9.6.4.2.1

Moment gradient from global analysis

Calculate buckling slenderness 3/9.7

$\lambda_{LT} = 0$ 3/9.7

Calculate limiting moment of resistance M_R 3/9.8

$M_R = M_{ult}$ 3/9.8

Calculate limiting stresses in extreme fibres 3/9.10.1.1

Effects due to vertical loads, shrinkage and differential temperature, and settlement of supports 3/9.2.1 3/9.9.7

Calculate design stresses in extreme fibres (sum over stages*)

* Beams must be checked for each stage of construction as well as the in-service condition. Sum all the stages up to the stage being considered. In the early stages, the beams will usually be non-composite.

Stresses OK? No → Redesign; Yes

For each web panel

Calculate equivalent stress in web panel 3/9.11.3

$\sigma_e < \sigma_{yw}$? No → Redistribute up to 60% of web load ($t = 0.4t_w$) 3/9.5.4; Yes

Restraint conditions around edges of web panel 3/9.11.4.2

Calculate buckling coefficients and ratios 3/9.11.4.3

Interaction ratio < 1.0? 3/9.11.4.4 No → Redesign; Yes

Longitudinal stiffeners to each web panel

Calculate effective stiffener section and σ_{ls}
Calculate stiffener stress σ_{se} at 0.4a from end 3/9.11.5 3/9.5.2

$\sigma_{se} < \sigma_{ls}$? Yes → SATISFACTORY; No → Redesign longitudinal stiffeners

SATISFACTORY

Figure 8.3 *Beams with longitudinal web stiffeners at ULS*

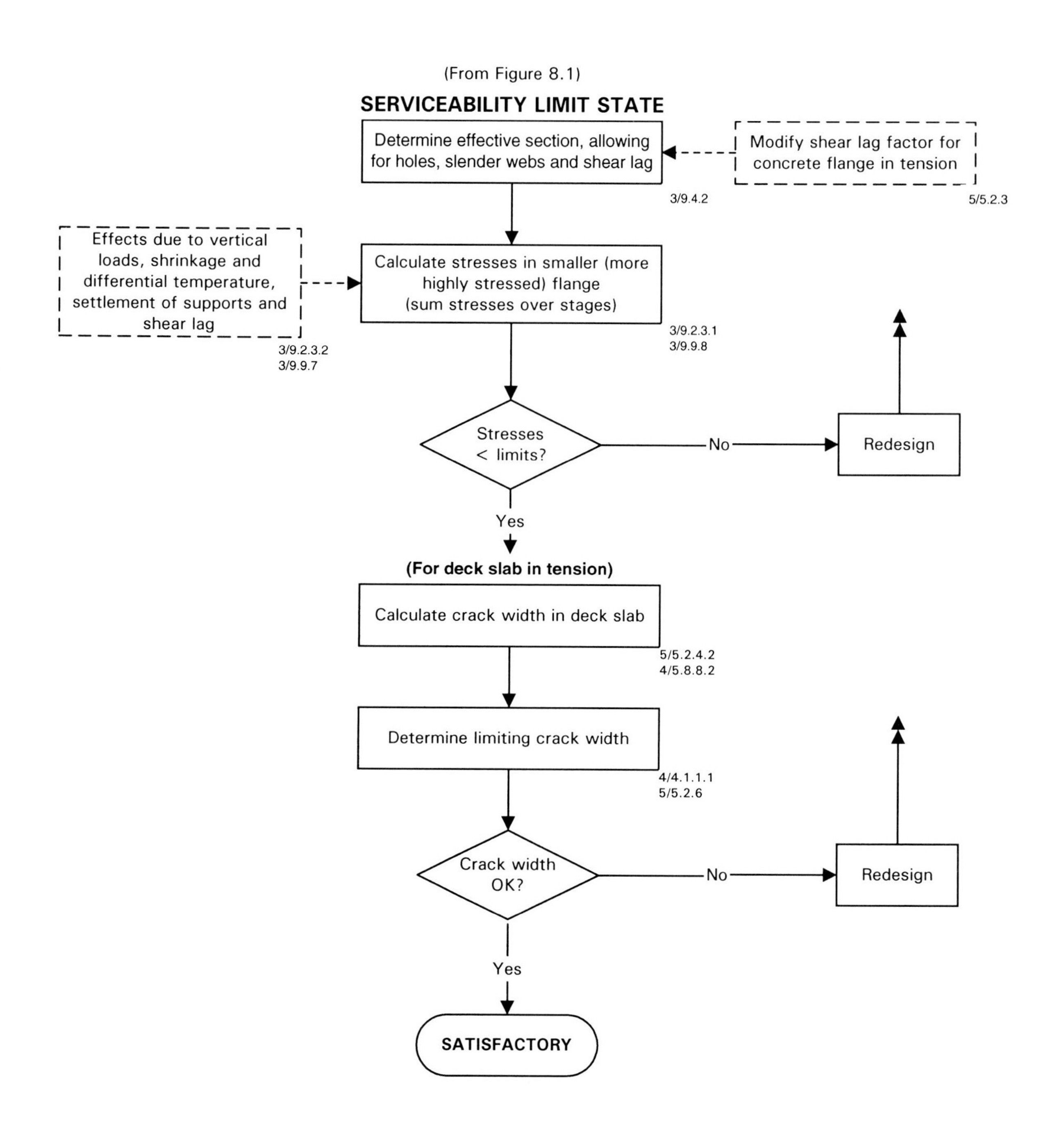

Figure 8.4 *Compact beams and non-compact beams with high shear lag at SLS*

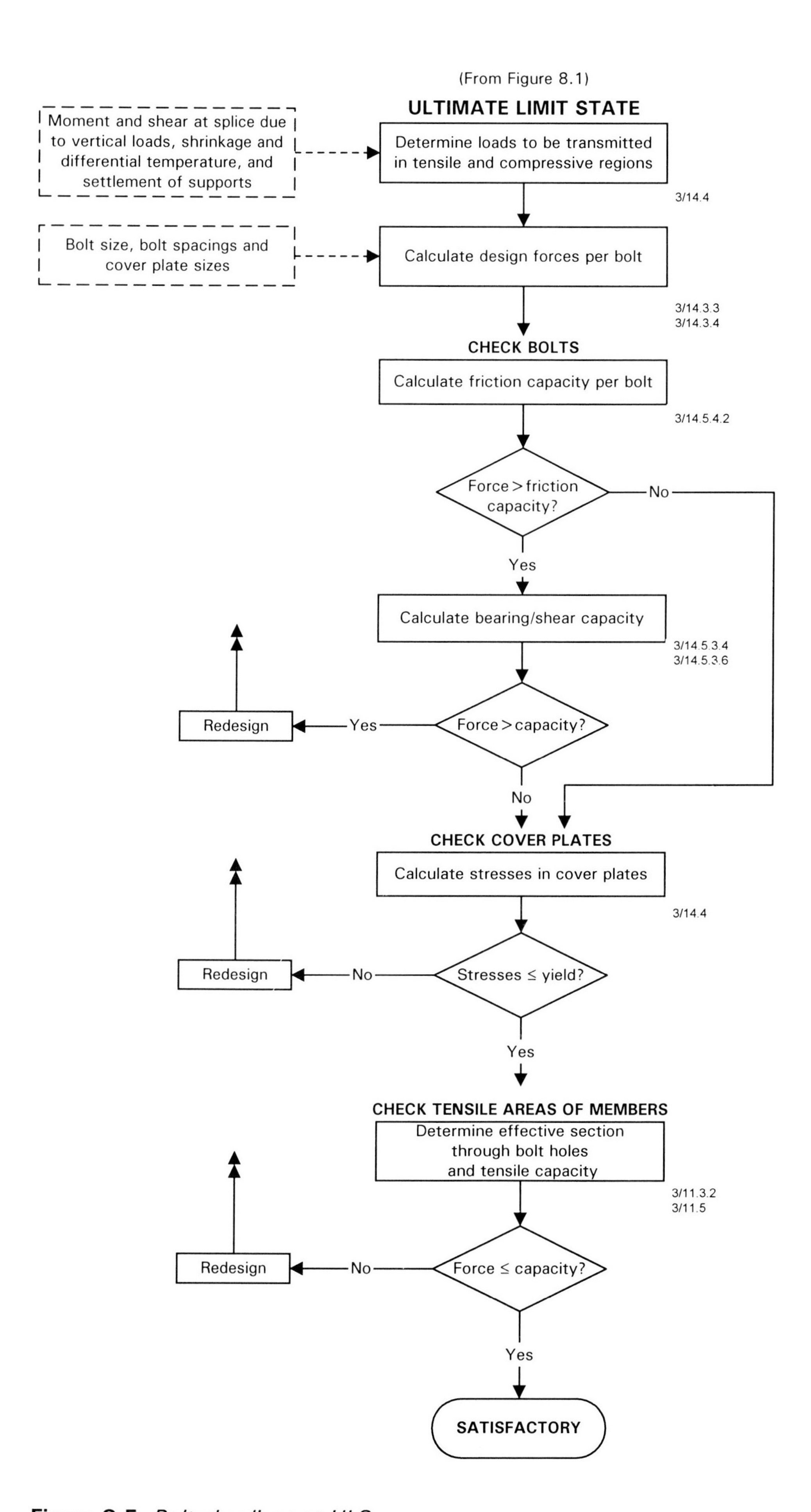

Figure 8.5 *Bolted splices at ULS*

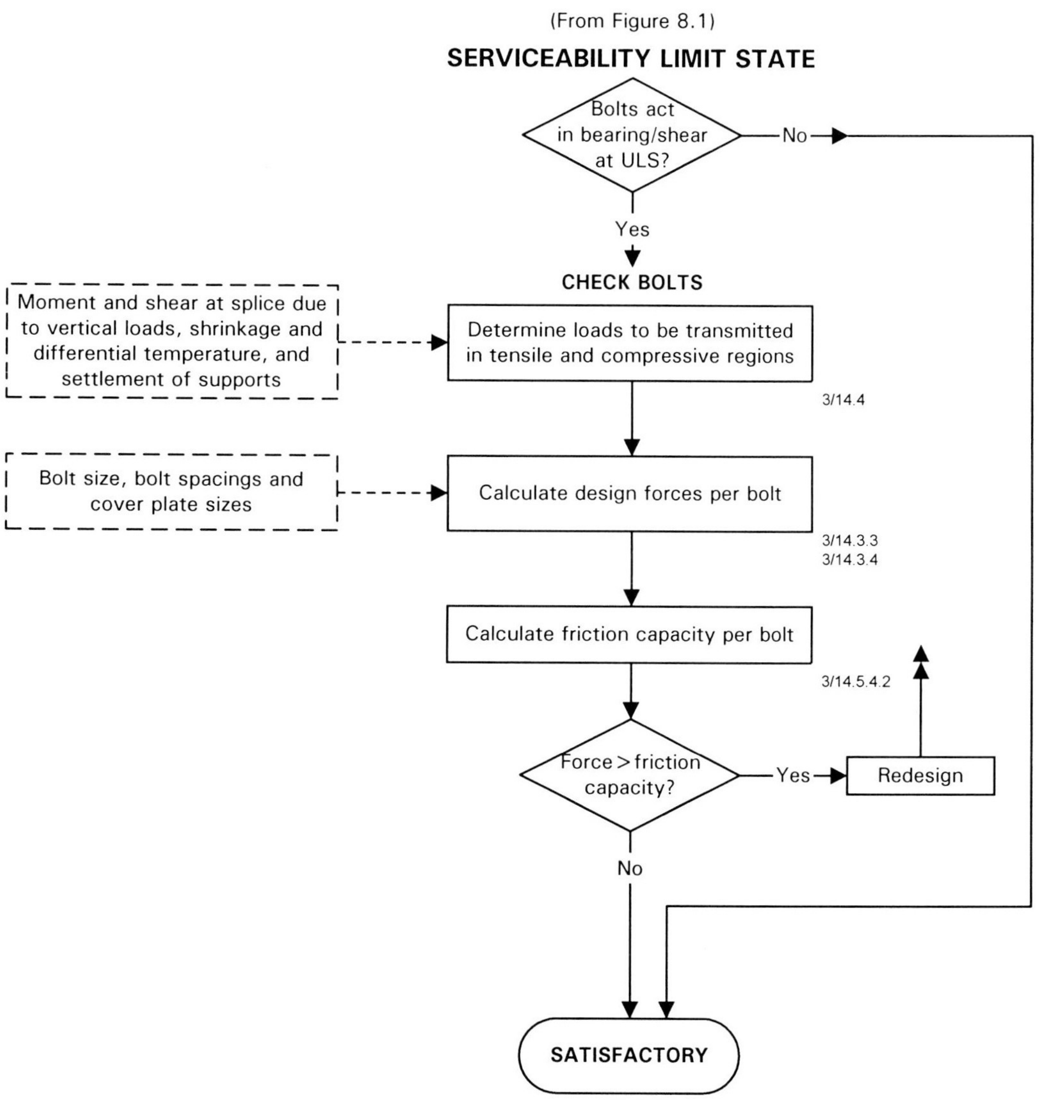

(From Figure 8.1)

SERVICEABILITY LIMIT STATE

Bolts act in bearing/shear at ULS?

No

Yes

CHECK BOLTS

Moment and shear at splice due to vertical loads, shrinkage and differential temperature, and settlement of supports

Determine loads to be transmitted in tensile and compressive regions

3/14.4

Bolt size, bolt spacings and cover plate sizes

Calculate design forces per bolt

3/14.3.3
3/14.3.4

Calculate friction capacity per bolt

3/14.5.4.2

Force > friction capacity?

Yes

Redesign

No

SATISFACTORY

Figure 8.6 *Bolted splices at SLS*

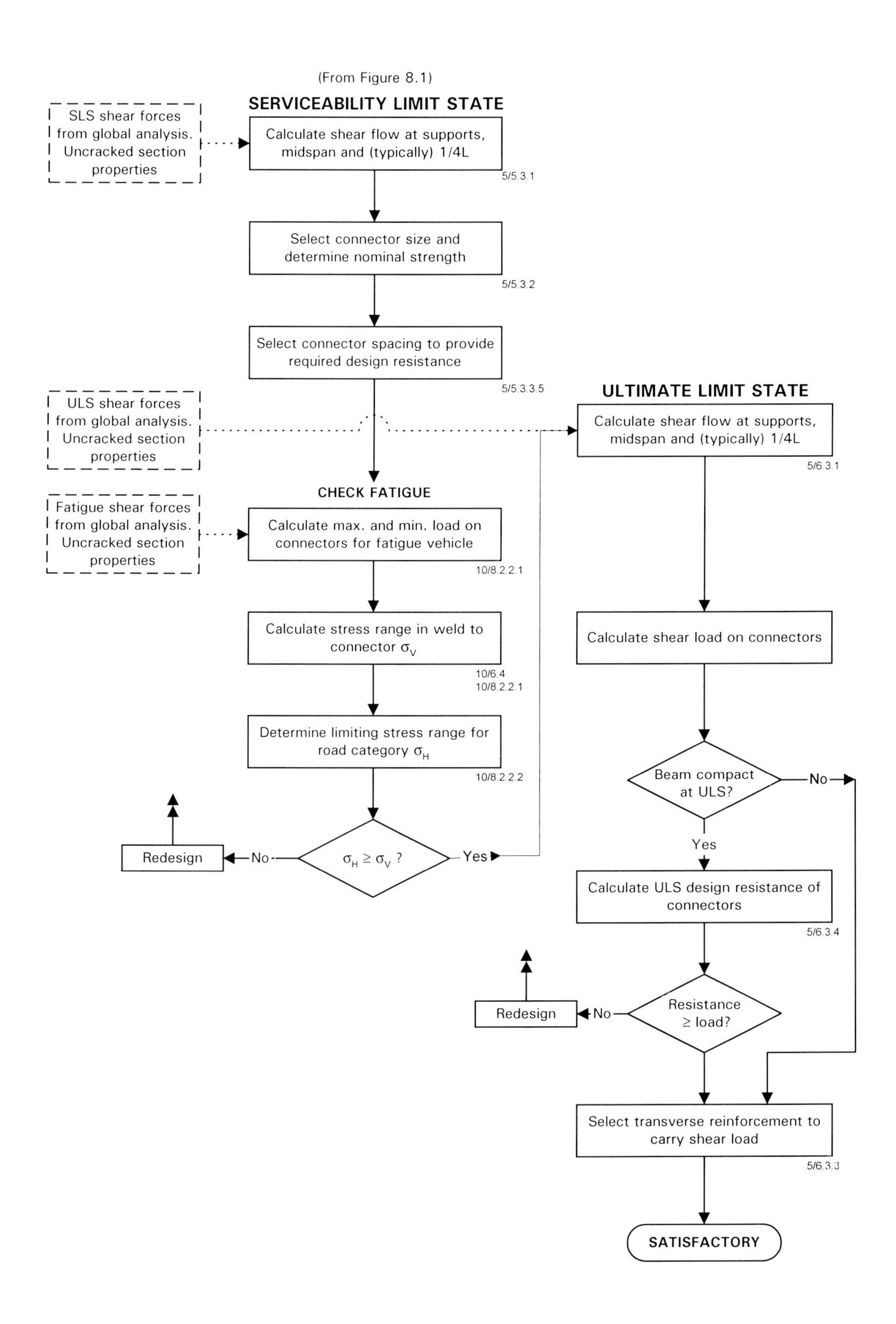

Figure 8.7 *Longitudinal shear connection*

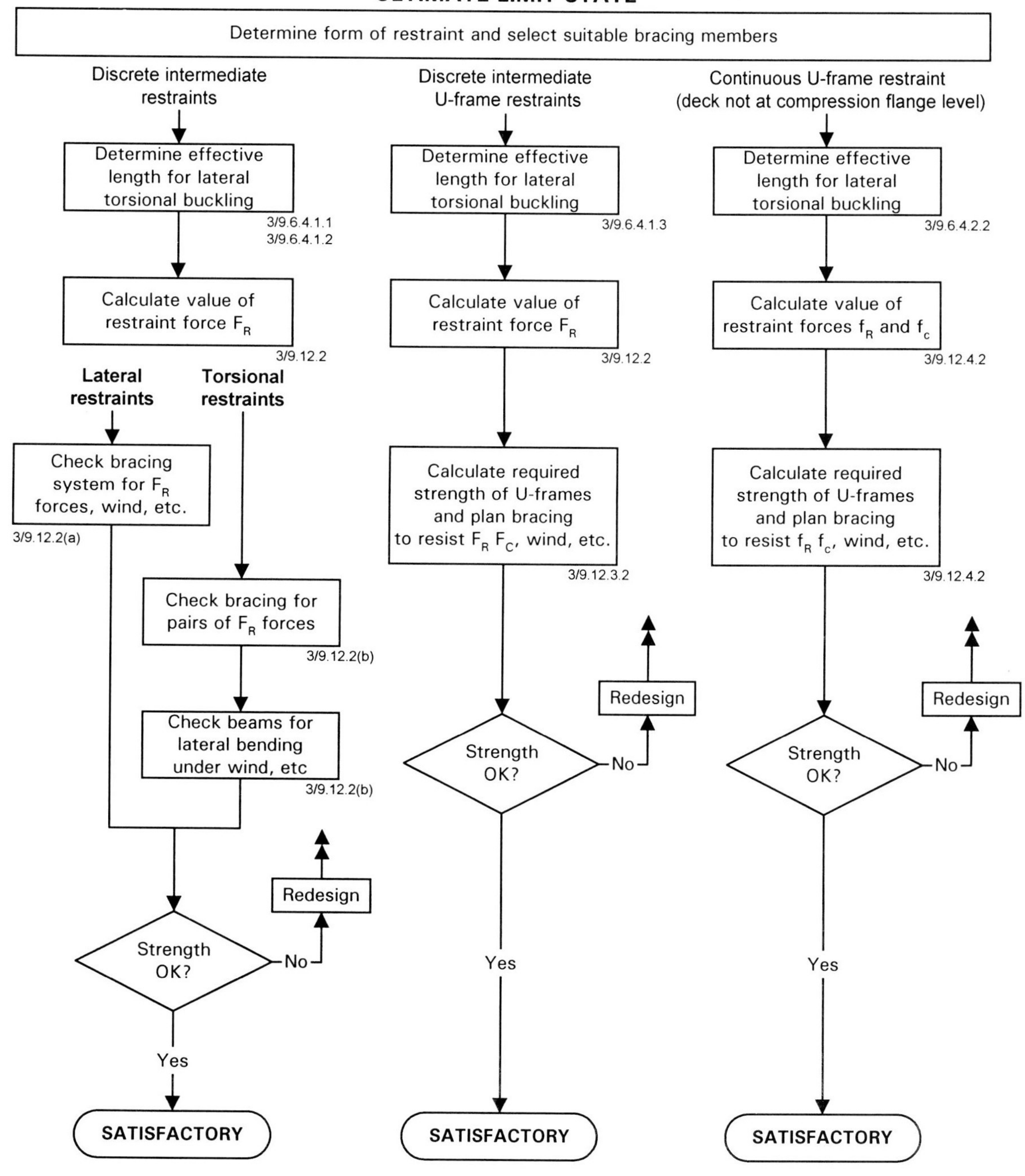

Figure 8.8 *Restraints to compression flanges*

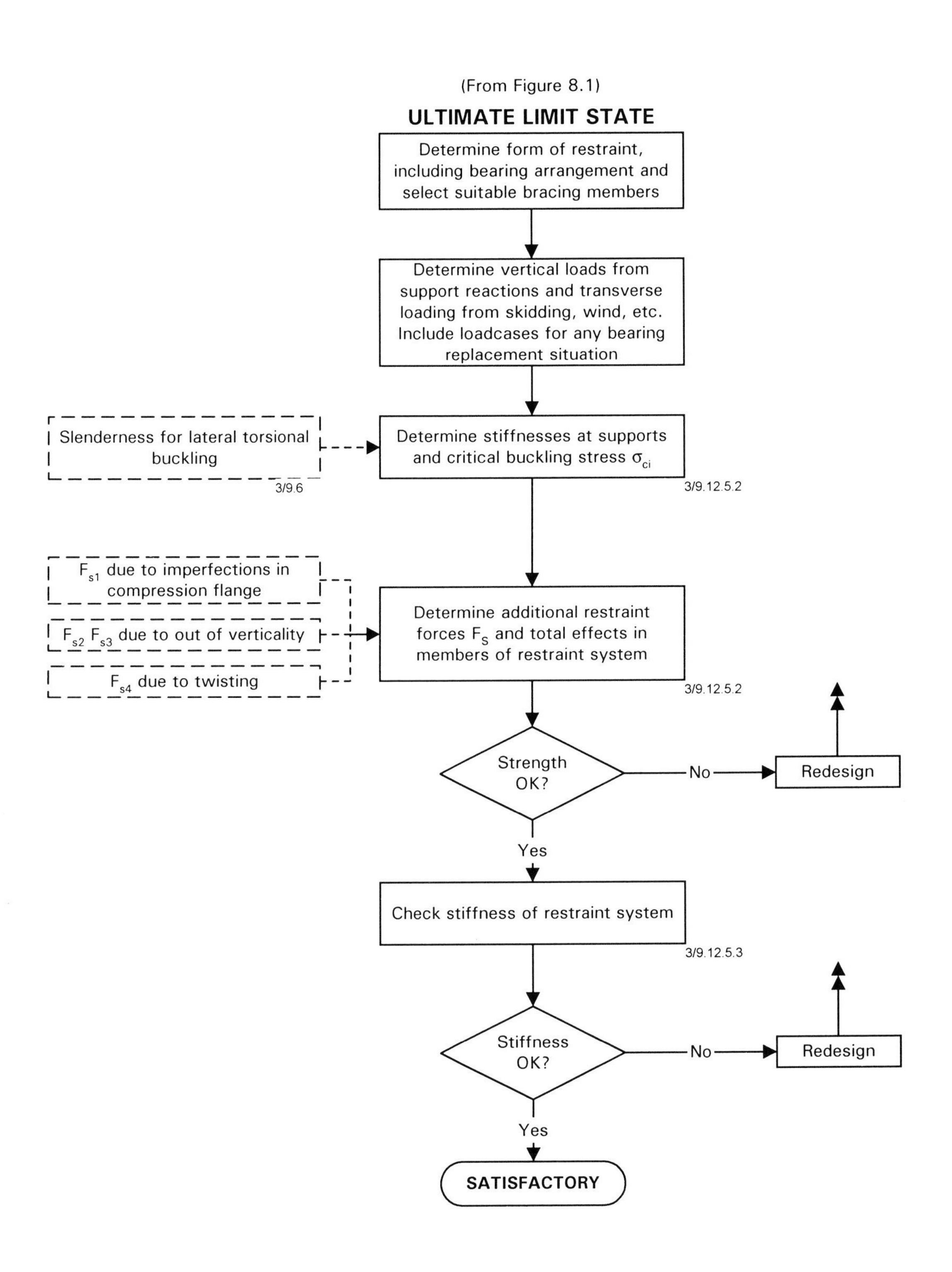

Figure 8.9 *Restraints at supports*

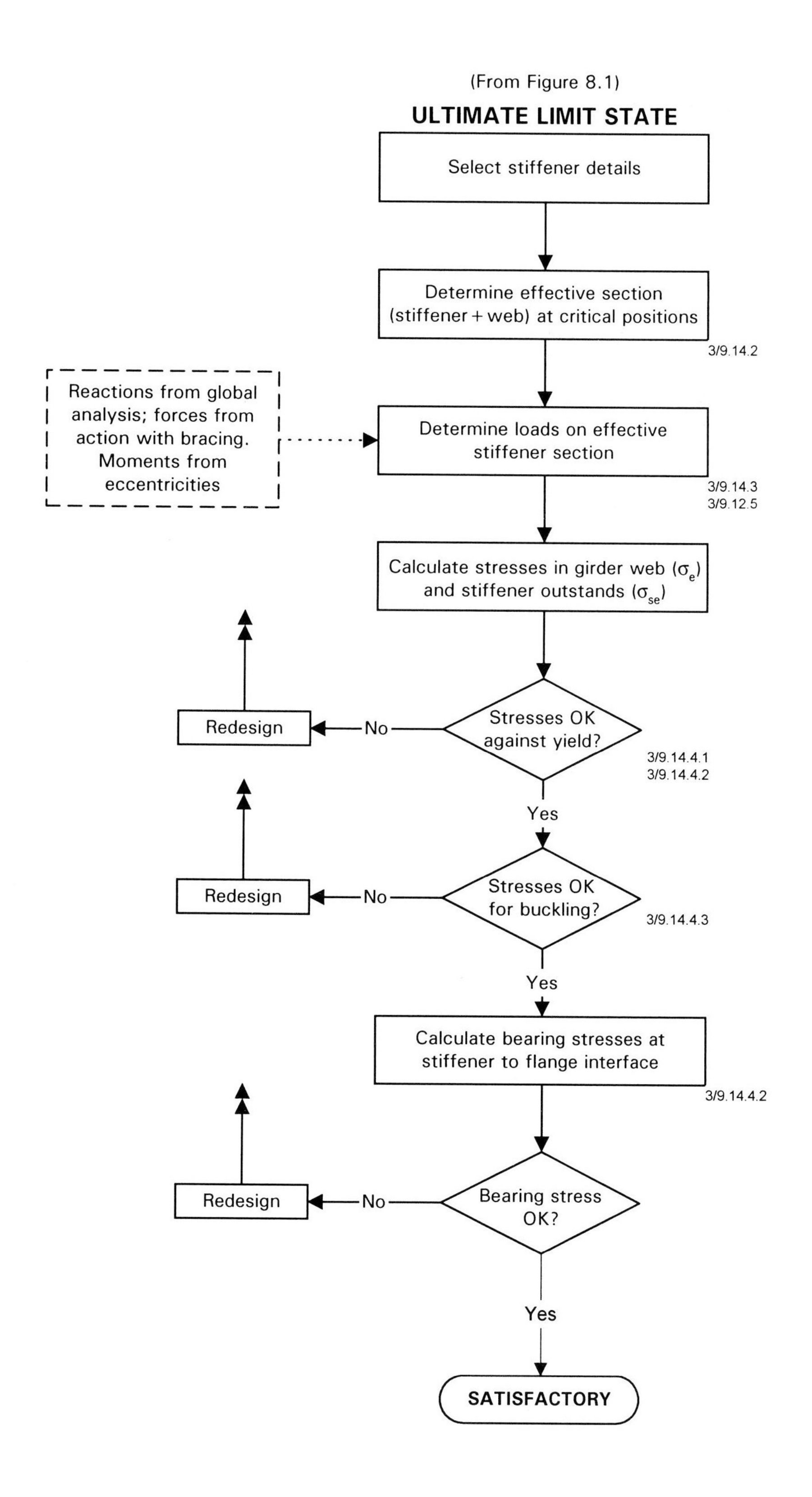

Figure 8.10 *Bearing stiffeners*

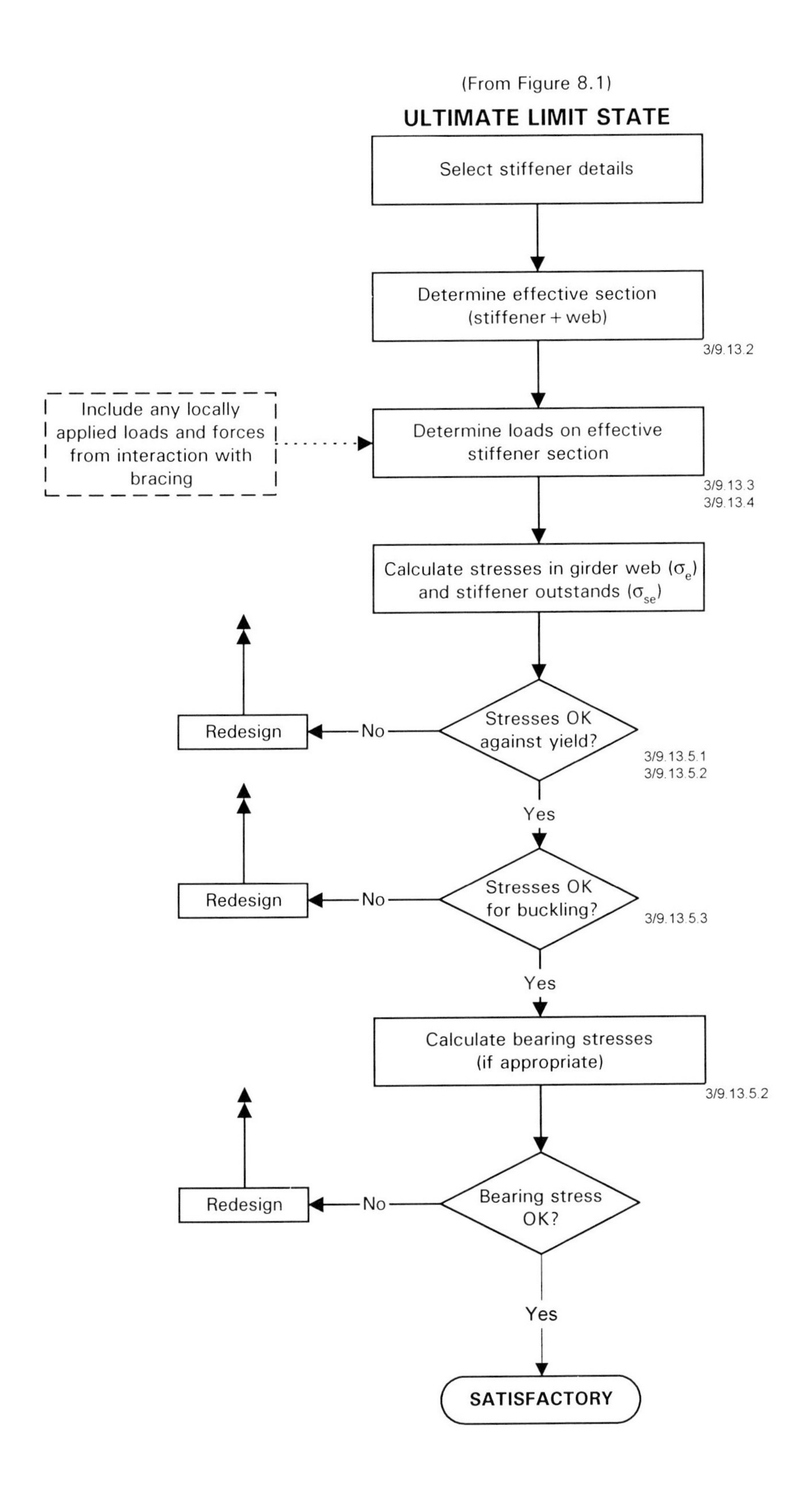

Figure 8.11 *Intermediate transverse web stiffeners*

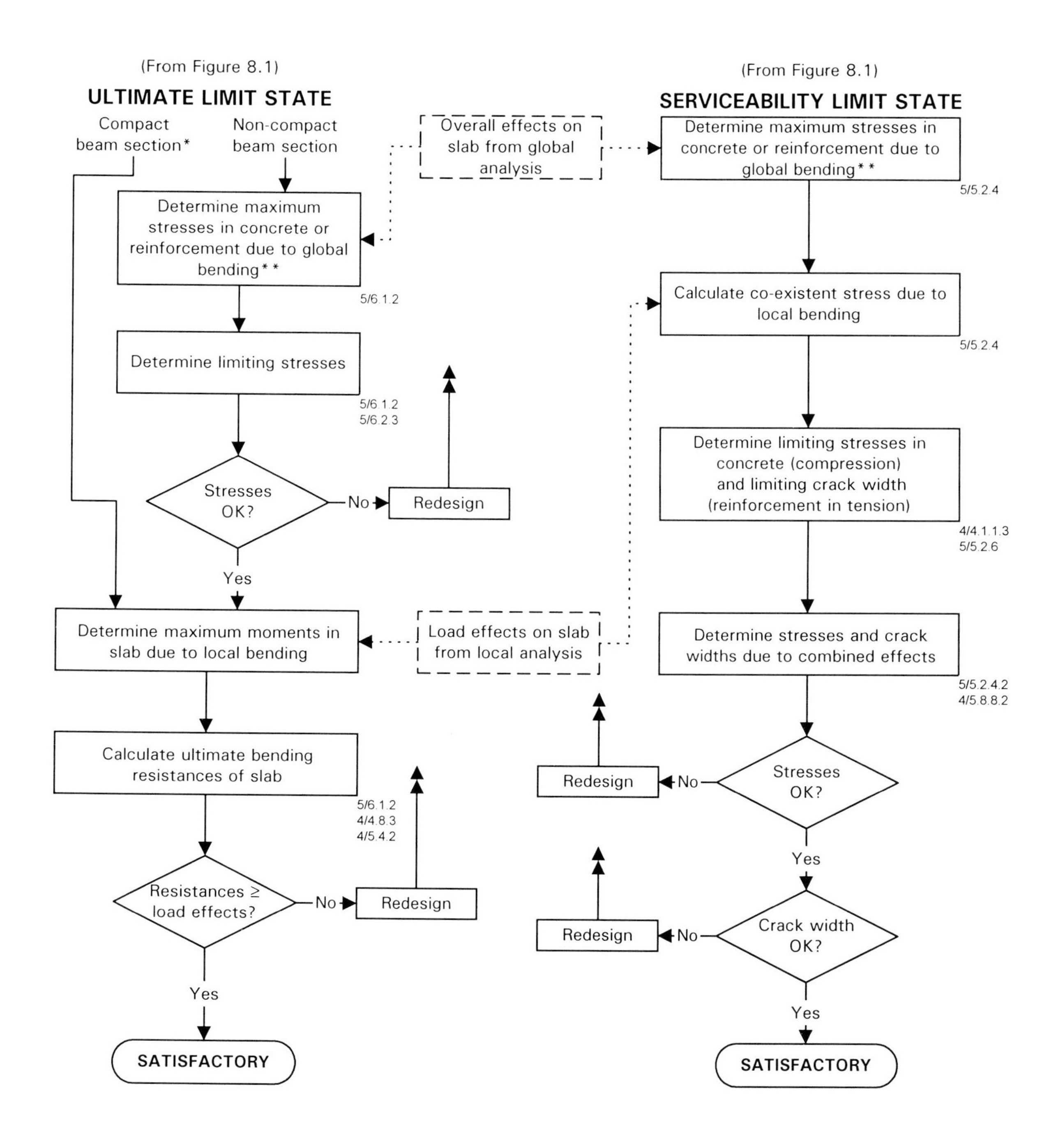

* Slabs forming part of a compact beam are verified for global bending at ULS as part of the beam. See Figure 8.2.

** Stresses determined from summation for beams built in stages

Figure 8.12 *Deck slab*

9 REFERENCES

1. Steel highway bridges protection guide
British Steel, 1997

2. LEE, D.J.
Bridge bearings and expansion joints (2nd Edition)
E & F SPON, 1994

3. BROWN, C.W. and ILES, D.C. (Editors)
Commentary on BS 5400-3:2000, Code of practice for the design of steel bridges (SCI P295)
The Steel Construction Institute, 2000

4. EVANS, J.E. and ILES, D.C. (Editors)
Steel Bridge Group: Guidance notes on best practice in steel bridge construction (SCI P185)
The Steel Construction Institute, 2000

5. BRITISH STANDARDS INSTITUTION
BS 5400 Steel concrete and composite bridges
-1:1988 General statement
-2:1978 Specification for loads
-3:2000 Code of practice for design of steel bridges
-4:1990 Code of practice for design of concrete bridges
-5:1979 Code of practice for design of composite bridges
-6:1999 Specification for materials and workmanship, steel
-7:1978 Specification for materials and workmanship, concrete, reinforcement and prestressing tendons
-8:1978 Recommendations for materials and workmanship, concrete, reinforcement and prestressing tendons
-9:1983 Bridge bearings
-10:1980 Code of practice for fatigue

6. ILES, D.C. (Editor)
Specification of structural steelwork for bridges: A model Appendix 18/1 (SCI P170)
The Steel Construction Institute, 1996

7. BRITISH STANDARDS INSTITUTION
BS 4360:1990 Specification for weldable structural steels (Withdrawn)
BSI

8. BRITISH STANDARDS INSTITUTION
BS EN 10025:1993 Hot rolled products of non-alloy structural steels - Technical delivery conditions
BSI

9. BRITISH STANDARDS INSTITUTION
BS EN 10113 Hot rolled products in weldable fine grain structural steels (3 Parts)
BSI, 1993

10. BRITISH STANDARDS INSTITUTION
BS EN 10155: 1993 Structural steels with improved atmospheric corrosion resistance - Technical delivery conditions
BSI

11. BRITISH STANDARDS INSTITUTION
BS 4:1993 Structural steel sections - Specification for hot-rolled sections
BSI

12. ILES, D.C.
Design guide for composite box girder bridges (SCI P140)
The Steel Construction Institute, 1994

13. BIDDLE, A.R., ILES, D.C and YANDZIO, E.
Integral steel bridges: Design guidance (SCI P163)
The Steel Construction Institute,1997

14. WAY, J.A. and YANDZIO, E.
Integral steel bridges: Design of a single span bridge - Worked example (SCI P180)
The Steel Construction Institute, 1997

15. WAY, J.A. and BIDDLE, A.R.
Integral steel bridges: Design of a multi-span bridge - Worked example (SCI P250)
The Steel Construction Institute, 1998

16. HAYWARD, A.C.G.
Composite steel highway bridges
British Steel, 1997

17. Structural sections to BS 4: Part 1 and BS 4848: Part 4 (Brochure)
British Steel, 1999

18. Plate products range of sizes for plates and profiling slabs (Brochure)
British Steel, 1999

19. BROWN, C.W. and ILES, D.C.
Design of steel bridges for durability (SCI P154)
The Steel Construction Institute, 1995

20. WEST, R.
Recommendations on the use of grillage analysis for slab and pseudo-slab bridge decks
Cement and Concrete Association and Construction Industry Research and Information Association, 1973

21. HAMBLY, E.C.
Bridge deck behaviour
E & F N Spon, 1991

22. JOHNSON, R.P. and HUANG, D.
Composite bridge beams with mixed-class cross sections
Structural Engineering International 2/95, IABSE, 1995

23. BRITISH STANDARDS INSTITUTION
BS 4395 Specification for high strength friction grip bolts and associated nuts and washers for structural engineering
-1:1969 General grade
-2:1969 Higher grade bolts and nuts and general grade washers
BSI

24. BRITISH STANDARDS INSTITUTION
BS 4604 Specification for the use of high strength friction grip bolts in structural steelwork. Metric series
-1:1970 General grade
-2:1970 Higher grade (parallel shank)
BSI

25. BRITISH STANDARDS INSTITUTION
BS 5135:1984 Specification for arc welding of carbon and carbon manganese steels
BSI

26. MATTHEWS, S.J.
Bearings and joints
Chapter 28 in the Steel Designers Manual (5th Edition)
Blackwell Scientific Publications, 1992

27. WESTERGAARD, H.M.
Computation of stresses in bridge slab due to wheel loads
Public Roads, Vol. 11, March 1930

28. PUCHER, A.
Influence surfaces of elastic plates
Springer-Verlag, Vienna, 1964

29. WOOD, R.H.
The reinforcement of slab in accordance with a predetermined field of moments
Concrete, Vol. 2, No. 2, February 1968

30. ARMER, G.S.T.
Discussion on reference 30.
Concrete, Vol. 2, No. 8, August 1968

APPENDIX A Guidance Notes

The following is a list of all the Guidance Notes in the SCI publication *Steel Bridge Group: Guidance notes on best practice in steel bridge construction* at the time of the second issue. These Notes provide additional advice to that presented here.

GN	Title
Section 1	Design – general
1.01	Glossary
1.02	Skew bridges
1.03	Bracing systems
1.04	Bridge articulation
1.06	Permanent formwork
1.07	Use of weather resistant steel
1.08	Box girder bridges
1.09	Comparison of bolted and welded splices
1.10	Half through bridges
Section 2	Design – detailing
2.01	Main girder make-up
2.02	Main girder connections
2.03	Bracing and cross-beam connections
2.04	Bearing stiffeners
2.05	Intermediate transverse web stiffeners
2.06	Connections made with HSFG bolts
2.07	Welds – how to specify
2.08	Attachment of bearings
Section 3	Materials and products
3.01	Structural steels
3.02	Through thickness properties
3.03	Bridge bearings
3.04	Welding processes and consumables
3.05	Surface defects on steel materials
3.06	Internal defects in steel materials
3.07	Specifying steel material
Section 4	Contract documentation
4.01	Drawings
4.02	Weld procedure trials
4.03	Allowing for permanent deformations
4.04	Alternative construction sequences

APPENDIX B Highways Authorities and Railtrack Documents

B.1 Highways Authorities

The requirements of the four Overseeing Organisations for highways [The Highways Agency (for England), The Scottish Executive Development Department (for Scotland), The National Assembly for Wales (for Wales) and The Department for Regional Development (for Northern Ireland)] are presented in two series of documents: the *Design Manual for Roads and Bridges* and the *Manual of Contract Documents for Highway Works*.

The *Design Manual for Roads and Bridges* (DMRB) comprises a set of Standards and Advice Notes on a variety of topics. Documents are now issued jointly on behalf of the four organisations, but the DMRB still includes some documents issued up to 1992 separately by the different authorities.

In the DMRB, the Standards and Advice Notes relating to the use of the various Parts of BS 5400, and to other matters relevant to composite bridge design, are as follows:

Document reference	Title	Date of issue	DMRB section
Standards and Advice Notes relating to Parts of BS 5400			
BD 15/92	General principles for the design and construction of bridges. Use of BS 5400: Part 1: 1988	Dec. 1992	1.3.2
BD 37/88	Loads for highway bridges (with revised Part 2 in an Appendix)	Aug. 1989	1.3
BD 13/90	Design of steel bridges. Use of BS 5400: Part 3: 1982	Feb. 1991	1.3
BA 19/85	The use of BS 5400: Part 3: 1982	Jan. 1985	1.3
BD 24/92	Design of concrete bridges. Use of BS 5400: Part 4: 1990	Nov. 1992	1.3.1
BD 16/82	Design of composite bridges. Use of BS 5400: Part 5: 1979	Nov. 1982	1.3
	Amendment no. 1	Dec. 1987	
	(see also the DTp combined document, Part 5 plus BD 16/82)		
BD 20/92	Bridge bearings. Use of BS 5400: Part 9: 1983	Oct. 1992	2.3.1
BD 9/81	Implementation of BS 5400: Part 10: 1980. Code of practice for fatigue	Dec. 1981	1.3
BA 9/81	The use of BS 5400: Part 10: 1980. Code of practice for fatigue	Dec. 1981	1.3
	Amendment no.1	Nov. 1983	
Other Standards and Advice Notes			
BD 7/81	Weathering steel for highway structures (revision due in 2001)	Aug. 1981	2.3
BD 28/87	Early thermal cracking of concrete	July 1987	1.3
	Amendment no.1	Aug. 1989	
BA 24/87	Early thermal cracking of concrete	July 1987	1.3
	Amendment no.1	Aug. 1989	
BA 36/90	The use of permanent formwork	Feb. 1991	2.3

BA 41/98	The design and appearance of bridges	Feb. 1998	1.3.11
BA 42/96	The design of integral bridges	Nov. 1996	1.3.12
BD 49/93	Design rules for aerodynamic effects on bridges	Jan. 1993	1.3.3
BD 52/93	The design of highway bridge parapets	Apr. 1993	2.3.3
BA 53/94	Bracing system and the use of U-frames in steel highway bridges	Dec. 1994	1.3.13
BD 57/95	Design for durability	Aug. 1995	1.3.7
BA 57/95	Design for durability	Aug. 1995	1.3.8
BD 60/94	Design of highway bridges for vehicle collision loads	Apr. 1994	1.3.5

The Manual of Contract Documents for Highway Works (MCDHW) comprises seven separate volumes of documents. In the MCDHW, certain sections of the *Volume 1: Specification for Highway Works* and *Volume 2: Notes for Guidance on the Specification for Highway Works* are relevant to bridge construction. Sections relevant to composite bridge construction are:

Series 1700	Structural concrete
Series 1800	Structural steelwork
Series 1900	Protection of steelwork against corrosion
Series 2100	Bridge bearings

B.2 Railtrack documents

Mandatory requirements are expressed in Group Standards. These are fairly short documents that state principles.

Group Standards

GC/RT5110 Design requirements for structures, Railtrack, 1996
GC/RT5112 Loading requirements for the design of bridges, Railtrack, 1997

Detailed requirements are expressed as recommendations in a Code of Practice.

Code of Practice

GC/RC5510 Recommendations for the design of bridges, Railtrack, 1998

APPENDIX C A guide to quick initial selection of web and flange sizes

The following is offered to designers with little previous experience of bridge design.

Governing shear will be seen when an HB vehicle of minimum length is placed directly over a girder adjacent to a support. If the girder is an edge girder, approximately 85% of the total shear will be carried by the girder under consideration. If it is an internal girder, a more appropriate proportion is 70%.

Governing moments usually occur when an HB vehicle is directly over the girder at midspan. Approximately 75% of the vehicle is carried by an edge girder; for an internal girder, the proportion is approximately 50%. Simple moment distribution will give realistic estimates of support and midspan moments.

Most designers size the web first so that it can carry 150% of the governing shear (the reserve is valuable in contributing to bending resistance). If the bottom flange is inclined, it will carry some of the shear and the web can be reduced in thickness accordingly.

The bottom flange is sized next to provide the required modulus.

At internal supports, the top flange is usually made half the area of the bottom, additional tensile capacity being provided by slab reinforcement.

At midspan, the top flange should only be reduced to 50% of the bottom if this will not create stability problems during construction. It will often be necessary to increase the top flange size for the construction conditions.

Worked Example Number 1

Contents

Commentary to calculation sheet

The worked example is a small four-span bridge carrying a two-lane road over a dual carriageway at a skew of about 33 °. Preliminary design has selected an arrangement of four main longitudinal girders acting compositely with an in-situ reinforced concrete slab. At the three intermediate supports each girder is supported on a separate column.

Silwood Park, Ascot, Berks SL5 7QN
Telephone: (01344) 623345
Fax: (01344) 622944

CALCULATION SHEET

1. OUTLINE DRAWINGS

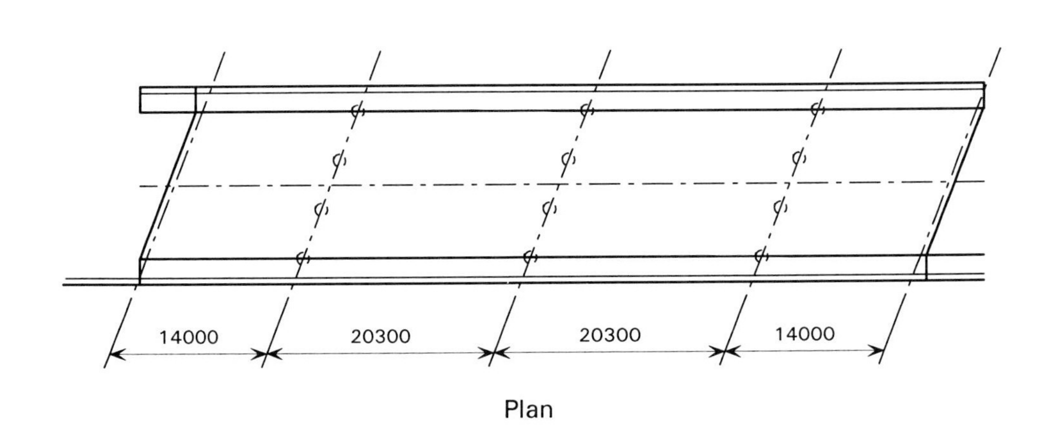

Plan

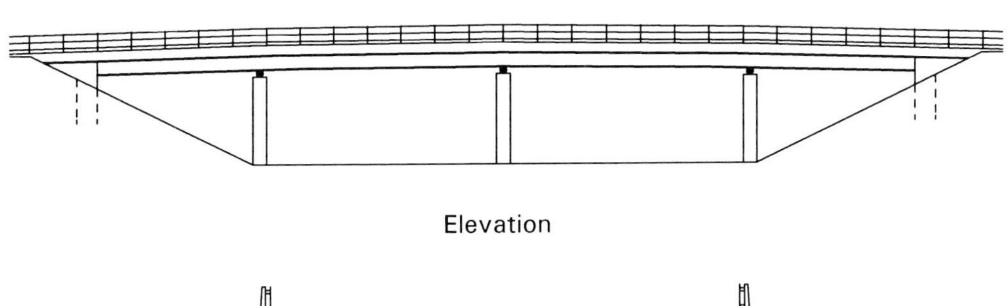

Elevation

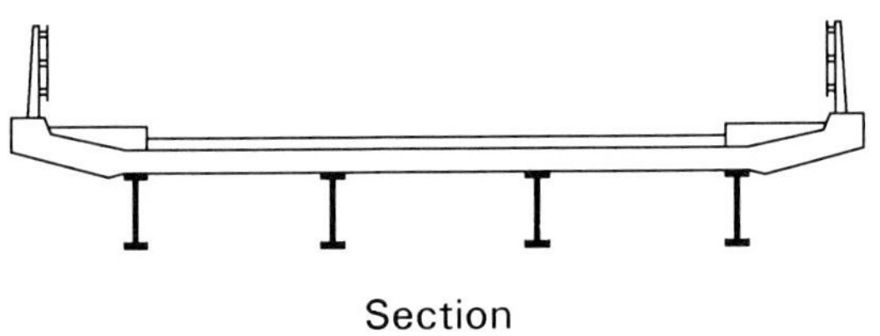

Section

Commentary to calculation sheet

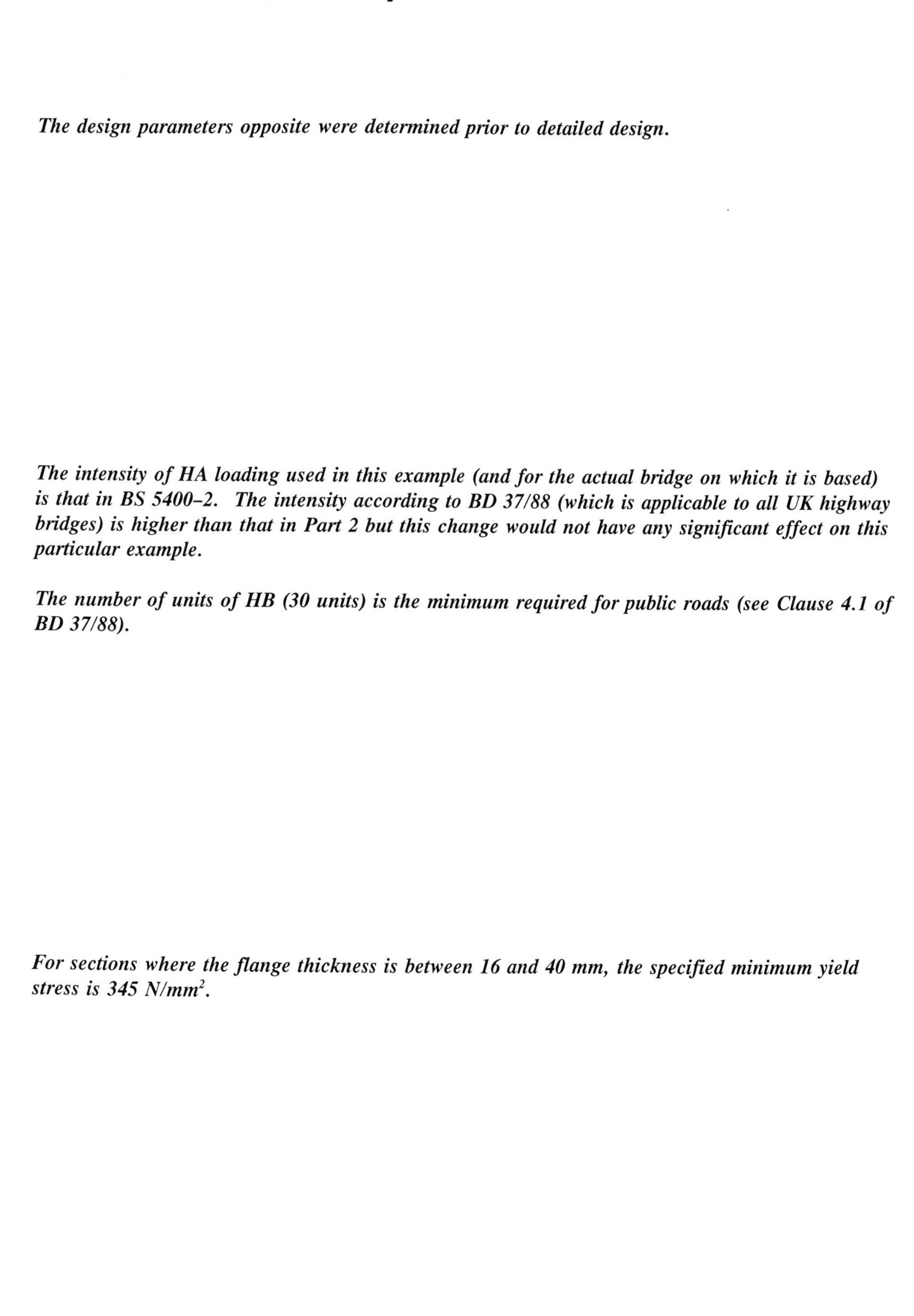

The design parameters opposite were determined prior to detailed design.

The intensity of HA loading used in this example (and for the actual bridge on which it is based) is that in BS 5400–2. The intensity according to BD 37/88 (which is applicable to all UK highway bridges) is higher than that in Part 2 but this change would not have any significant effect on this particular example.

The number of units of HB (30 units) is the minimum required for public roads (see Clause 4.1 of BD 37/88).

For sections where the flange thickness is between 16 and 40 mm, the specified minimum yield stress is 345 N/mm^2.

Silwood Park, Ascot, Berks SL5 7QN
Telephone: (01344) 623345
Fax: (01344) 622944

CALCULATION SHEET

Job No:	BCR825	Sheet 2 of 32	Rev B
Job Title	Design of composite bridges – Worked example no. 1		
Subject	Design Data		
Client	SCI	Made by DCI	Date Sep 2000
		Checked by NK	Date Dec 2000

2. DESIGN DATA

Spans = *14.0, 20.3, 20.3, 14.0 m*
Skew = *33°*
Carriageway = *7.3 m wide, 2 lanes*
Surfacing = *125 mm thick (including waterproofing)*
Footways = *1.5 m wide, each side*
Location = *Northern England*
Design life = *120 years*
Settlement = *up to 10 mm at any support*

Loading

Live load — *HA (2 lanes)*
HB (30 units)
Footway

Densities — *Concrete: 25 kN/m³*
Surfacing: 22 kN/m³

Minimum effective bridge temperature

Minimum shade air temp.		*-16 °C*
Minimum effective bridge temperature (Group 3)		*-14 °C*
Allow for 100 m above sea level		*-0.5 °C*
U_e	=	*-14.5 °C*

Design parameters

Assume Grade 40 concrete and Grade S355 steel.

σ_y = 355 N/mm²
f_{cu} = 40 N/mm²
E_s = 205000 N/mm²
E_{cs} = 31000 N/mm²
E_{cL} = 15500 N/mm²

$$\text{Short-term modular ratio } \alpha_e = \frac{205000}{31000}$$

$$= 6.61$$

Commentary to calculation sheet

For initial design, maximum moments at an intermediate support are based approximately on moments on a uniform continuous beam. For all spans loaded, the support moment is close to that for a fixed-ended span ($wL^2/12$). For UDL over two adjacent spans, the moment is slightly greater ($wL^2/10$). For a point load in one interior span, the maximum influence line coefficient is about 0.08L (or $M = WL/12$).

The Steel Construction Institute

Silwood Park, Ascot, Berks SL5 7QN
Telephone: (01344) 623345
Fax: (01344) 622944

3. *INITIAL DESIGN*

Overall width of deck 11.4 m
Assume deck slab thickness 235 mm
Assume 4 No. girders at 2.65 mm centres
Assume overall depth 1/20 of main span = 1015 mm
Girder depth = 1015 − 235 = 780 mm

Approximate design moment over central support

Dead load:	*Nominal load*	γ_{fL}	*ULS Load*
Girder ~ say 2.7 kN/m (1 kN/m²)	2.7	1.05	2.8
2.65 m width of slab	15.6	1.15	17.9
2.65 m width of surfacing	7.3	1.75	12.8
			33.5 kN/m

DL moment over central support

$$= \frac{wL^2}{12} \text{ (approx.)}$$

$$= \frac{33.5 \times 20.3^2}{12} = 1150 \text{ kNm}$$

Live load:
Use the greater of HA and HB loading

HA loaded length = 40.6 m
∴ *loading intensity* = 26.1 kN/m

$$\text{ULS load/beam} = \frac{26.1 \times 2.65 \times 1.5}{3.65} = 28.4 \text{ kN/m}$$

$$\text{KEL/beam} = \frac{120 \times 2.65 \times 1.5}{3.65} = 130.7 \text{ kN}$$

$$\text{HA moment} = \frac{wL^2}{10} \text{ (approx.)} = \frac{28.4 \times 20.3^2}{10} = \underline{1170 \text{ kNm}}$$

$$\text{KEL moment} = \frac{WL}{12} \text{ (approx.)} = \frac{130.7 \times 20.3}{12} = \underline{221 \text{ kNm}}$$

Load/beam from each HB axle
By static distribution

$$\frac{75\,(1.65 + 2.65 + 1.65 + 0.65)}{2.65} = 187 \text{ kN}$$

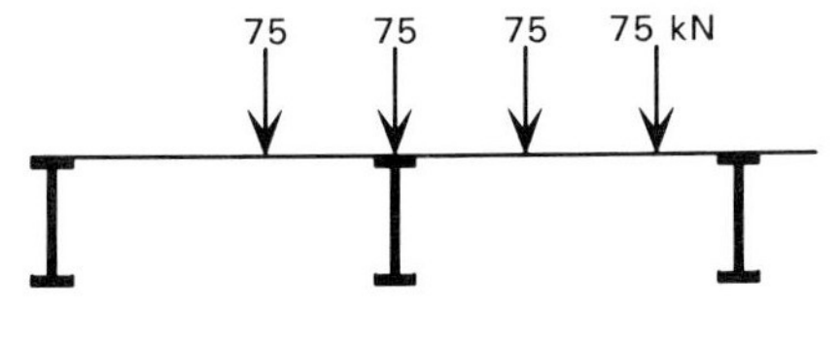

$$\text{Moment} \approx \frac{WL}{12} = \frac{4 \times 1.3 \times 187 \times 20.3}{12}$$

$$= 1644 \text{ kNm}$$

Dead + live load moments = 1150 + 1644 = 2794 kNm
Allow for shrinkage effects Design moment = 3000 kNm

Commentary to calculation sheet

It is assumed that the flanges for the beams will be between 16 and 40 mm thick, so the design yield stress will be 345 N/mm^2. Partial factors are γ_{f3} = 1.1 and γ_m = 1.05.

The factor of 0.9 in the evaluation of required plastic modulus is an assumed allowance for reduction for lateral torsional buckling slenderness.

In the example given in SCI publication P065, a Universal Beam section was used and the composite section over intermediate supports was just compact according to the rules then in BS 5400-3. Those rules considered the depth of web in compression in the elastic section. The amended rules for compact classification consider the depth of web in compression in the plastic section and this is more onerous; a composite section using a Universal Beam section will not be compact in hogging moment regions. The example has therefore been reworked using a fabricated section, with a thicker web than in a UB section.

The top flange reinforcement is about the minimum that will comply with crack width requirements at SLS. A larger size bar is used as bottom reinforcement to cater for local sagging moments.

For calculation of plastic modulus appropriate to the compression flange yield strength (here 345 N/mm^2), the reinforcement area is transformed to an equivalent area of the same yield strength, in accordance with Clause 5/6.2.2. Partial factor γ_m = 1.05 applied to structural steel yield strength and design strength of reinforcement taken as 0.87f_{ry}.

Note that if a larger area of reinforcement is used, the plastic neutral axis might rise so high in or above the web that the chosen section would no longer be compact.

The Steel Construction Institute	Job No: *BCR825*	Sheet *4* of *32*	Rev *B*
	Job Title *Design of composite bridges – Worked example no. 1*		
Silwood Park, Ascot, Berks SL5 7QN Telephone: (01344) 623345 Fax: (01344) 622944	Subject *Initial design*		
	Client *SCI*	Made by *DCI*	Date *Sep 2000*
CALCULATION SHEET		Checked by *NK*	Date *Dec 2000*

Minimum required plastic modulus $= \dfrac{3000 \times 10^6}{(345/(1.1 \times 1.05)} \times \dfrac{1}{0.9}$

$= 11.2 \times 10^6 \text{ mm}^3$

If this is provided by a couple at say 850 mm centres, the required area at bottom flange level is given by:

$A = 11.2 \times 10^6 / 850 = 13200 \text{ mm}^2$

Web thickness for compact section over support

If, say, $m = 0.85$

Limiting web depth according to 3/9.3.7.2 is

$374 \times (355 / 345)^{0.5} / (13 \times 0.85 - 1) = 37.7\, t_w$

Web depth will be about 720 mm (780 – 2 × 30, say), so web thickness needed = 20 mm

Half the area of web in compression is effectively at bottom flange level for plastic moment

Area $= 0.85 \times 720 \times 20 \times ½ = 6120 \text{ mm}^2$

So required flange area is about $13200 - 6120 = 7080 \text{ mm}^2$

Try flange 300 × 25 mm (outstand = 140 mm = $5.6t_{fo}$ *– OK)*

Area $= 7500 \text{ mm}^2$

Use a symmetric section (flanges the same size). Web depth = 780 – 50 = 730 mm

Provide reinforcement to balance the web area assumed to be effectively in compression

Reinforcement ($\sigma_{ry} = 460 \text{ N/mm}^2$)

Reinforcement area required:

$(0.85 - 0.15) \times 720 \times 20 = 10220 \text{ mm}^2$ *(structural steel units)*

$= 10220 \times (345 / 1.05) / (460 \times 0.87)$

$= 8390 \text{ mm}^2$ *(actual area)*

Try T16 (bottom) + T12 (top), both at 100 mm centres

Area $= \pi (8^2 + 6^2) \times 2650 / 100 = 8325 \text{ mm}^2$

Should be sufficient for strength

Commentary to calculation sheet

Global analysis may be carried out entirely on a grillage model, or the dead load analysis may be carried out on a simple line-beam model and the live load analysis on a grillage model.

A line beam model was used for calculation of moments and shears due to dead loads in this example. Intermediate node positions were chosen to be the same as the grillage nodes, to allow easy correlation of dead and live load moments and shears at critical positions.

The grillage shown opposite was used for live load analysis. The UDL and KEL loads in the lanes and on the footways were converted into series of point loads on the members. It was not considered necessary to represent the cantilevers.

In the complete design, moments and shears were determined for inner and outer beams. In this example, only the figures for the inner beam are presented.

Transverse global moments were used in the design of the deck slab, which was reinforced orthogonally to the main beams. The design of the deck slab is not included in this example.

The Steel Construction Institute

Silwood Park, Ascot, Berks SL5 7QN
Telephone: (01344) 623345
Fax: (01344) 622944

CALCULATION SHEET

Job No:	BCR825		Sheet	5 of 32	Rev	B
Job Title	Design of composite bridges – Worked example no. 1					
Subject	Grillage model					
Client	SCI	Made by	DCI		Date	Sep 2000
		Checked by	NK		Date	Dec 2000

4. GRILLAGE MODEL

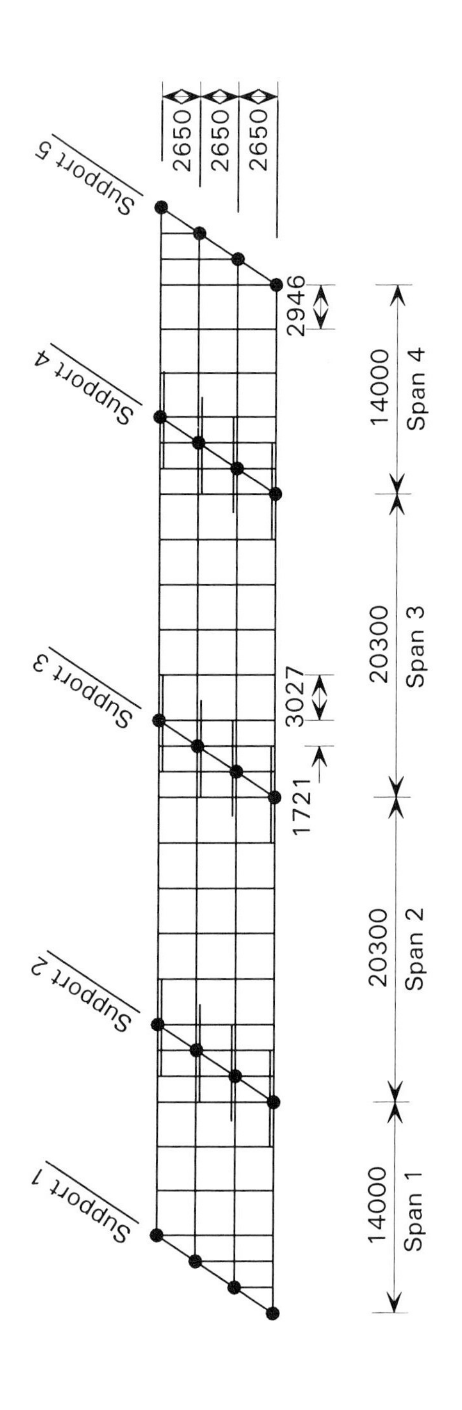

Layout of grillage model

Note:
All longitudinal elements represent composite beam and slab
Longitudinal elements shown thus: ═══ have cracked section properties
All transverse elements represent a width of slab only

Commentary to calculation sheet

Calculations for the outer beam will be similar, though the geometry of the cantilever will lead to different values.

The area of the reinforcement that is effective in the elastic section is reduced by 2½% from gross area on account of its lower modulus (200 N/mm^2 cf. 205 N/mm^2).

Silwood Park, Ascot, Berks SL5 7QN
Telephone: (01344) 623345
Fax: (01344) 622944

5. SECTION PROPERTIES (Inner beam)

(a) Elastic properties

For bare steel section

	A	y	Ay	Ay^2	I
	mm^2	mm	$10^3\ mm^3$	$10^6\ mm^4$	$10^6\ mm^4$
Top flange 300×25	7500	7675	5756	4418	0.4
Web 20×730	14600	390	5694	2221	648
Bottom flange 300×25	7500	12.5	94	1	0.4
	29600		11544	6640	649

$\bar{y} \quad = \quad 390\ mm$

$I \quad = \quad 649 + 6640 - 11544 \times 0.390 \quad = \quad 2.787 \times 10^9\ mm^4$

__Sagging regions__

(Short-term section)

Equivalent width of slab (short-term modulus)

$$= \frac{2650}{6.61} = 401\ mm$$

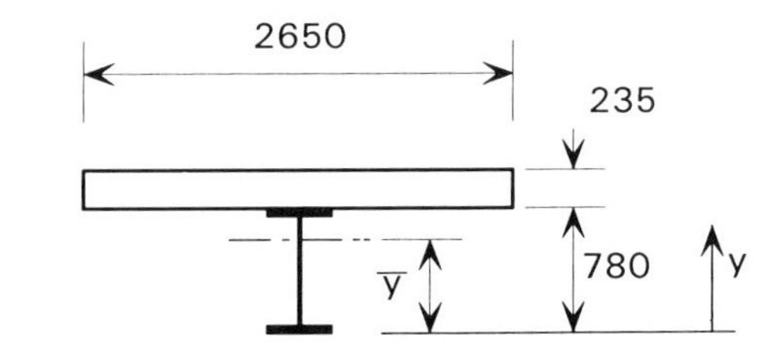

	A	y	Ay	Ay^2	I
	mm^2	mm	$10^3\ mm^3$	$10^6\ mm^4$	$10^6\ mm^4$
Slab 401 × 235	94235	897.5	84576	75907	434
Beam	29600		11544	6640	649
	123835		96120	82547	1083

$\bar{y} \quad = \quad 776.2\ mm$ *(in top flange)*

$I \quad = \quad (82547 + 1083 - 96120 \times 0.7762) \times 10^6 \quad = \quad 9.022 \times 10^9\ mm^4$

Similar calculations for long-term section give: $I \quad = \quad 7.681 \times 10^9\ mm^4$

__Hogging regions__

Effective area of T12

$= 113.1 \times 200 / 205 = 110.3\ mm^2$

Effective area of T16

$= 201.1 \times 200 / 205 = 196.2\ mm^2$

26 T12 at 100
26 T16 at 100
96
73
780

	A	y	Ay	Ay^2	I
	mm^2	mm	$10^3\ mm^3$	$10^6\ mm^4$	$10^6\ mm^4$
Top bars (26.5 T12)	2924	949	2775	2633	
Bottom bars (26.5 T16)	5199	853	4434	3782	
Beam	29600		11544	6640	649
	37723		18753	13055	649

$\bar{y} \quad = \quad 497.1\ mm$

$I \quad = \quad (649 + 13055 - 18753 \times 0.4971) \times 10^6 \quad = \quad 4.382 \times 10^9\ mm^4$

Commentary to calculation sheet

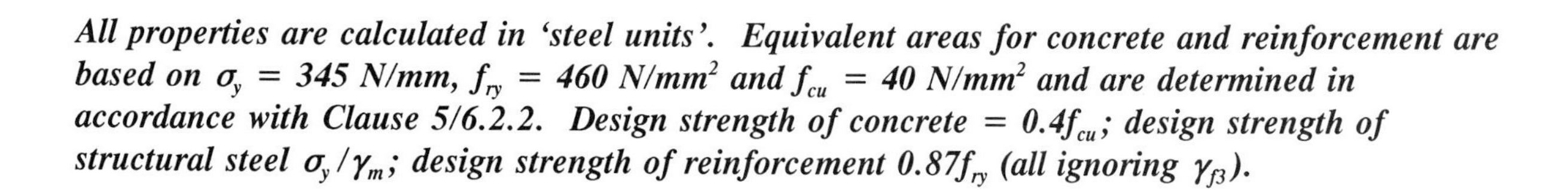

All properties are calculated in 'steel units'. Equivalent areas for concrete and reinforcement are based on σ_y = 345 N/mm, f_{ry} = 460 N/mm^2 and f_{cu} = 40 N/mm^2 and are determined in accordance with Clause 5/6.2.2. Design strength of concrete = 0.4f_{cu}; design strength of structural steel σ_y / γ_m; design strength of reinforcement 0.87f_{ry} (all ignoring γ_{f3}).

The movement from the assumed neutral axis = (unbalanced area) / 2 × web thickness. (The stress in that area changes from compressive to tensile yield.)

The beam is compact at the supports (in hogging) and compact in midspan. The load effects due to shrinkage, differential temperature and creep must be taken into account at both locations if the slenderness is greater than the limiting value given by 3/9.2.1.3.

The movement from the assumed neutral axis = (unbalanced area) / slab width. (Equivalent width = 129 mm in steel units.)
The stress in that area changes from compressive strength to zero.

(b) PLASTIC SECTION PROPERTIES and CLASSIFICATION

Bare steel beam

	A (mm^2)	y (mm)	Ay ($10^3 mm^3$)
Top flange	7500	377.5	2831
Top half web	7300	182.5	1332
Bottom half web	-7300	-182.5	1332
Bottom flange	-7500	-377.5	2831
	0		8326

$Z_{pe} = 8.33 \times 10^6 \, mm^3$

Over centre support (hogging)

Area ratio $= 0.87 \times 460 \times 1.05 / 345 = 1.218$

Try plastic neutral axis (PNA) at top of web

	A (mm^2)	y (mm)	Ay ($10^3 mm^3$)
Top T12 (26.5×113×1.218)	3650	949	3464
Bottom T16 (26.5×201×1.218)	6490	853	5536
Top flange	7500	767.5	5756
Web	-14600	390	-5694
Bottom flange	-7500	12.5	-94
	-4460		8968
Move PNA down by 111.5	4460	699.3	3119
			12087

$Z_{pe} = 12.09 \times 10^6 \, mm^3$

$y_{pl} = 643.5 \, mm$

Depth of web in compression $= 730 - 111.5 = 618.5 \, mm \quad \therefore m = 618.5 / 730 = 0.847$

Limiting depth $= 374 t_w / (13m - 1) = 374 \times 20 / (13 \times 0.847 - 1) = 747 \, mm$

$\therefore$ *Section is compact*

Midspan (sagging)

Area ratio $= (0.4 \times 40) / (345 / 1.05) = 0.0487$

Try PNA at top of beam

	A (mm^2)	y (mm)	Ay ($10^3 mm^3$)
Slab 2650 × 235 × 0.0487	-30330	897.5	-27221
Top flange	7500	767.5	5756
Web	14600	390.0	5694
Bottom flange	7500	12.5	94
	-730		-15677
Move PNA up by 5.6 (into slab)	730	782.8	571
			-15106

$Z_{pe} = 15.11 \times 10^6 \, mm^3$ *NA is in slab, so section is compact*

Commentary to calculation sheet

As explained in SCI P289, half of the simple torsional inertia is attributed to each orthogonal direction.

Torsional inertia of the appropriate widths of slab, calculated by the $bt^3/6$ rule, may be used in the grillage model for all slab elements, including those in the cracked zone over intermediate supports.

Torsional inertia of rectangular section depends on the ratio of long to short sides. A simple tabular expression is given by West [1].

---

[1] *WEST, R. Recommendations on the use of grillage analysis for slab and pseudo-slab bridge decks, Cement and Concrete Association and Construction Industry Research and Information Association, 1973*

Properties for transverse elements

Slab alone:

Bending inertia $I = \dfrac{bd^3}{12\,\alpha_e}$ $(\alpha_e = \text{modular ratio} = 6.6)$

Torsional inertia $J = \dfrac{bd^3}{6\,\alpha_e}$

Several transverse elements are required, each representing a different width of slab, as shown in layout (Sheet 5)

b (mm)	*d* (mm)	*I* (mm^4)	*J* (mm^4)
1721	235	0.282×10^9	0.564×10^9
2374	235	0.389×10^9	0.778×10^9
2946	235	0.483×10^9	0.965×10^9
3027	235	0.496×10^9	0.992×10^9

End diaphragm is a reinforced concrete beam 700 × 770 mm

J for a rectangle given by $\dfrac{k_1 \,.\, b^3 \,.\, d}{\alpha_e}$ *(when* $b < d$*)*

For $\dfrac{d}{b} = \dfrac{770}{700} = 1.1,\ k_1 = 0.154$

Hence, $J = 616 \times 10^9\ \text{mm}^4$

and $I = 4.035 \times 10^9$

Torsional inertia of longitudinal elements (inner beam)

J for 2650 width of slab $= 0.868 \times 10^9\ \text{mm}^4$

Contribution from I section is negligible

Commentary to calculation sheet

Case 1 could be subdivided to reflect the progressive erection of steel beams, but when the splices are close to the points of contraflexure, this refinement is not usually necessary.

Extra ULS models may be required for critical stages of erection – for example to check moments on the beam in one of the midspan regions under the weight of wet concrete.

Silwood Park, Ascot, Berks SL5 7QN
Telephone: (01344) 623345
Fax: (01344) 622944

CALCULATION SHEET

Job No:	BCR825	Sheet	9 of 32	Rev	B
Job Title	Design of composite bridges – Worked example no. 1				
Subject	Loads				
Client	SCI	Made by	DCI	Date	Sep 2000
		Checked by	NK	Date	Dec 2000

6. LOADS

Dead loads:

Self weight of beam	*0.0296 × 77*	=	*2.28 kN/m*
*Concrete slab**	*2.65 × (0.235 + 0.025) × 25*	=	*17.23 kN/m*
Surfacing	*2.65 × 0.125 × 22*	=	*7.29 kN/m*

**Concrete slab includes 25 mm allowance for possible concreting to underside of flanges*

Apply loads for a succession of structures:

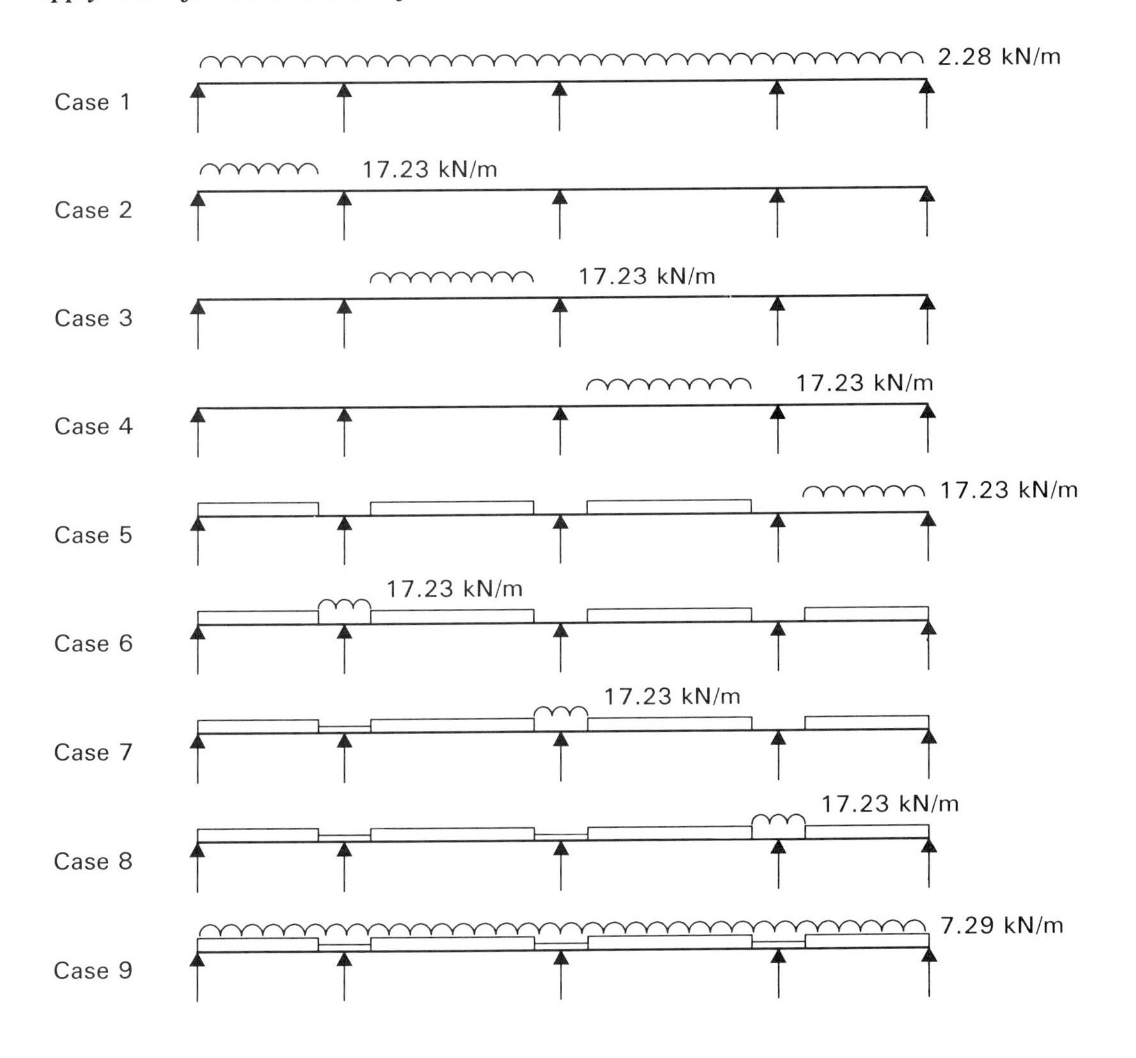

Commentary to calculation sheet

Highway loading, Clause 2/6; footway loading, Clause 2/7.

Selection of appropriate lengths for the HB vehicle to achieve maximum effect should use influence surfaces appropriate to the grillage model. It is usually quite adequate to use the simple influence lines that are more readily available from a line-beam model.

Knife edge loads are positioned within the area of HA distributed loading to cause maximum effects.

Primary effects

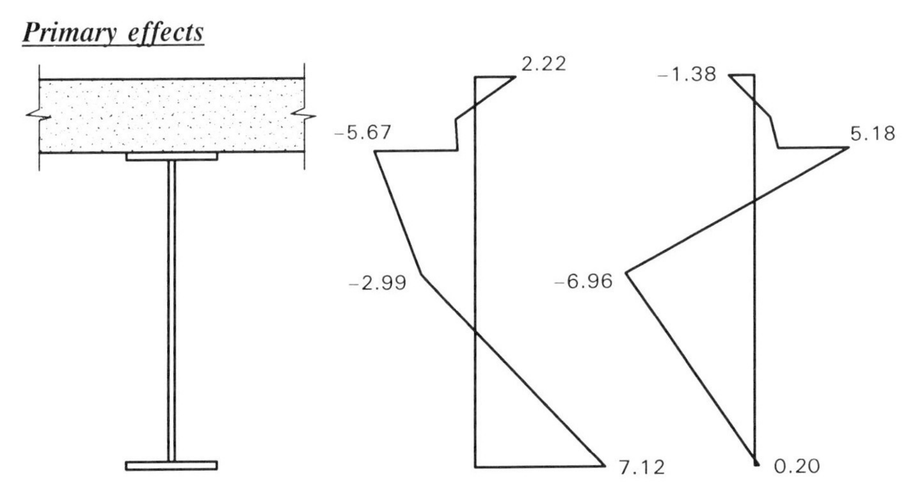

Positive difference Negative difference
(Negative difference stresses calculated in a similar manner)

Secondary effects

From line beam analysis, with moments applied at the ends of the beams:
For positive differential temperature, BMD is 240 plus { –240 / +41 / – 16 }
For negative differential temperature, BMD is 170 plus { –170 / +29 / – 16 }

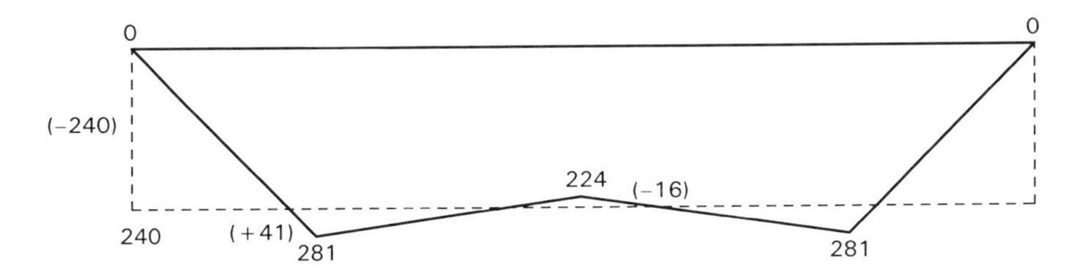

Positive difference

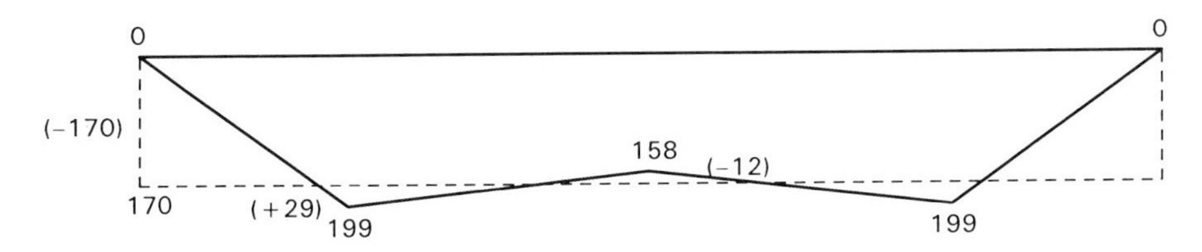

Negative difference

Commentary to calculation sheet

Values of E_{cs} and ϕ_c are given in Clause 5/5.4.3 for concrete that is within the usual range of mix proportions, as defined by Figure 5 of Part 5.

Partial factors for shrinkage effects are given in 5/4.1.2 as 1.2 at ULS and 1.0 at SLS.

Shear lag and cracking of concrete in tension should be ignored, for primary effects, as for differential temperature (see Clause 5/5.4.2.1).

Primary stresses calculated as on Sheet 11:

$$\sigma = \frac{-P_t}{A} + \frac{-M_t y}{I} + E\varepsilon$$

Primary stresses and secondary effects arising from the requirements of continuity are to be taken into account in accordance with Clause 3/9.9.7. [They cannot be neglected, because the beam is not compact throughout its length, nor is $\lambda_{LT} \leq 30$ (Clause 3/9.2.1.3).]

See Clause 5/5.4.2.2 for coefficients of expansion for concrete.

Silwood Park, Ascot, Berks SL5 7QN
Telephone: (01344) 623345
Fax: (01344) 622944

CALCULATION SHEET

Job No: *BCR825*	Sheet *13* of *32*	Rev *B*
Job Title *Design of composite bridges – Worked example no. 1*		
Subject *Loads*		
Client *SCI*	Made by *DCI*	Date *Sep 2000*
	Checked by *NK*	Date *Dec 2000*

Shrinkage modified by creep

$\epsilon_{cs} \quad = \quad -200 \times 10^{-6}$ *(open air conditions)*

$\phi_c \quad = \quad 0.4$

$$\alpha_e = \frac{E_s}{\phi_c E_c} = \frac{205}{0.4 \times 31} = 16.53$$

Long-term properties for creep

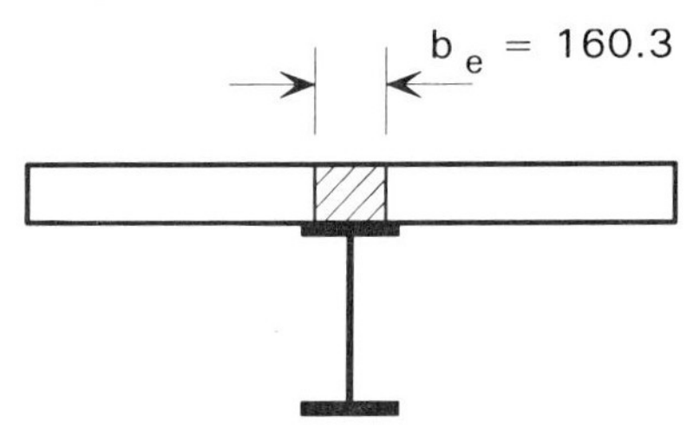

$$\text{Effective breadth} = \frac{2650}{16.53} = 160.3 \text{ mm}$$

(steel units)

$A = 67200 \text{ mm}^2 \qquad I = 7.225 \times 10^9 \text{ mm}^4 \qquad y = 674 \text{ mm}$

Force to restrain shrinkage $= A_c \times \phi E_c \times \epsilon_{cs}$

$= 2650 \times 235 \times 12400 \times -200 \times 10^{-6} = -1544 \text{ kN}$

Moment $= -1544 \times 10^3 \times 217 = -335 \text{ kNm}$

Stresses at top and botttom of slab, top and bottom of girder are:
-1.72, -12.95, 28.05, -9.22 N/mm² respectively (all in 'steel units')
And in reinforcement: 36.1 and 31.5 N/mm²

Secondary effects

From line beam analysis, with moments applied at the ends, and using long-term section properties.

For shrinkage, BMD is –335 plus { +335 / –62 / +26 }

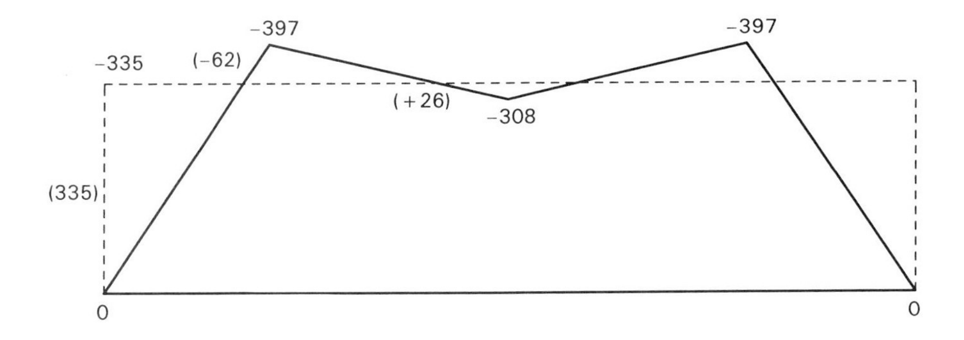

Commentary to calculation sheet

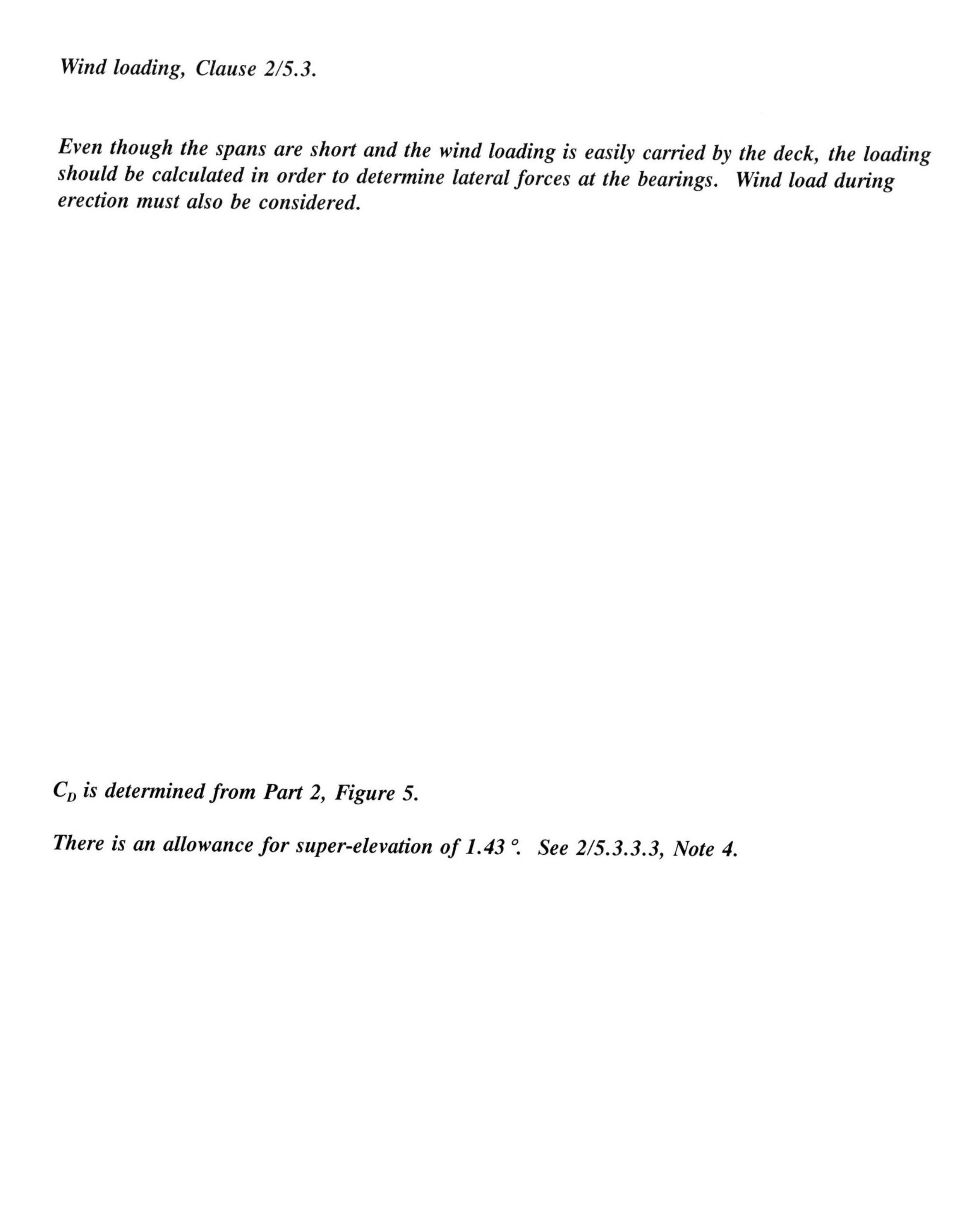

Wind loading, Clause 2/5.3.

Even though the spans are short and the wind loading is easily carried by the deck, the loading should be calculated in order to determine lateral forces at the bearings. Wind load during erection must also be considered.

C_D is determined from Part 2, Figure 5.

There is an allowance for super-elevation of 1.43 °. See 2/5.3.3.3, Note 4.

The Steel Construction Institute

Silwood Park, Ascot, Berks SL5 7QN
Telephone: (01344) 623345
Fax: (01344) 622944

CALCULATION SHEET

Wind loading

Mean hourly wind speed $v = 31$ m/s

Consider 3 conditions:

PERMANENT CONDITIONS		*TEMPORARY CONDITION (No live load)*
No live load	*With live load*	
$K_1 = 1.0$ $S_1 = 1.0$		$K_1 = 0.85$ $S_1 = 1.0$
Gust factor for $H_t = 8$ m $S_2 = 1.52$ (LL ≤ 20 m) or 1.45 (LL ≤ 60 m)		*Gust factor* $S_2 = 1.52$
Max. gust speed $v_c = 31 \times 1.0 \times 1.0 \times 1.52$ $= 47.1$ m/s (LL ≤ 20 m) $v_c = 44.95$ m/s (LL ≤ 60 m)	*Max. gust speed with LL* $v_c = 35$ m/s	$v_c = 40.05$ m/s
Min. gust speed $k_2 = 0.96$ $v_c' = 31 \times 0.96 \times 1.0$ $= 29.8$ m/s	*Min. gust speed* $v_c' = 35 \times 0.96/1.45$ $= 23.2$ m/s	
2.40	$D_i = 2.50$ 1.14	0.24 0.78
Solid area $A_1 = 2.40$ m³/m $b/d = 11.4/2.40 = 4.75$ $C_D = 1.36\,[1+(0.03 \times 1.43)]$ $= 1.418$	$d_3 = 3.64$ m For C_D, $d = d_c = 2.5$ $b/d = 11.4/2.5 = 4.56$ $C_D = 1.36\,[1+(0.03 \times 1.43)]$ $= 1.418$	$d = 1.02$ $b/d = 8.25/1.02 = 8.09$ $C_D = 1.29$
$P_t = q\,A_1\,C_D$ $= 0.613\,v_c^2 \times 2.40 \times 1.418$ $= 2.086\,v_c^2$		
$P_t = 4.63$ kN/m ($L = 20$ m) $= 4.21$ kN/m ($L = 60$ m)	$P_t = 0.613\,v_c^2 \times 3.64 \times 1.418$ $= 3.88$ kN/m	$P_t = 0.613\,v_c^2 \times 1.02 \times 1.29$ $= 1.29$ kN/m

Commentary to calculation sheet

Values of γ_{fL} *for settlement are taken from BD 37/88.*

Live load factors are for combination 1.

The Steel Construction Institute

Silwood Park, Ascot, Berks SL5 7QN
Telephone: (01344) 623345
Fax: (01344) 622944

CALCULATION SHEET

Job No: *BCR825*	Sheet *15* of *32*	Rev *B*
Job Title *Design of composite bridges – Worked example no. 1*		
Subject *Results of global analysis*		
Client *SCI*	Made by *DCI*	Date *Sep 2000*
	Checked by *NK*	Date *Dec 2000*

7. RESULTS OF GLOBAL ANALYSIS

Inner beam over centre support

	Loading	Nominal		ULS			SLS		
				γ_{fL}	BM	Shear	γ_{fL}	BM	Shear
1	Self weight UB	-86	25	1.05	-90	26	1.0	-86	25
2	Slab on span 1	+42	-3	1.15	+48	-3	1.0	+42	-3
3	Slab on span 2	-296	+19	1.15	-340	+22	1.0	-296	+19
4	Slab on span 3	-372	+118	1.15	-428	+136	1.0	-372	+118
5	Slab on span 4	+33	-11	1.15	+38	-13	1.0	+33	-11
6	Slab over support 2	-5	0	1.15	-6	0	1.0	-5	0
7	Slab over support 3	-73	+58	1.15	-84	+67	1.0	-73	+58
8	Slab over support 4	-8	+2	1.15	-9	+2	1.0	-8	+2
9	Superimposed load	-234	+77	1.75	-410	+135	1.2	-281	+92
					-1281	372		-1046	300
	HB 30 (11 m, for max. BM)	-547	+192	1.3	-711	+250	1.1	-602	+211
	HA (lane 2, spans 2/3)	-370	+125	1.3	-481	+163	1.1	-407	+138
	FW (next lane 2, spans 2/3)	-20	-2	1.3	-26	-3	1.0	-20	-2
	Settlement, support 2 (10 mm)	-69	+4	1.2	-83	+5	1.0	-69	+4
	Settlement, support 4 (10 mm)	-69	+10	1.2	-83	+12	1.0	-69	+10
					-2665	+799		-2213	661
	HB 30 (6m, for max. shear)	-402	+393	1.3	-523	+511			
	HA (lane 2, spans 2/3)	-370	+125	1.3	-481	+163			
	Settlement, supports 2 and 4				-166	+17			
					-2451	1063			

Midspan (span 3–4)

	Loading	Nominal		ULS			SLS		
				γ_{fL}	BM	Shear	γ_{fL}	BM	Shear
1	Self weight UB	41		1.05	+43		1.0	+41	
2	Slab on span 1	+16		1.15	+18		1.0	+16	
3	Slab on span 2	-117		1.15	-135		1.0	-117	
4	Slab on span 3	+427		1.15	+491		1.0	+427	
5	Slab on span 4	-70		1.15	-81		1.0	-70	
6	Slab over support 2	-12		1.15	-14		1.0	-12	
7	Slab over support 3	+13		1.15	+15		1.0	+13	
8	Slab over support 4	+11		1.15	+13		1.0	+11	
9	Superimposed load	+168		1.75	+294		1.2	+202	
					644			511	
	HB 30 (11 m, for max. BM)	+737		1.3	+958		1.1	+811	
	HA (lane 2, spans 2/3)	+318		1.3	+413		1.1	+350	
	FW (next lane 2, spans 2/3)	+48		1.3	+62		1.0	+48	
	Settlement, support 3 (10 mm)	+19		1.2	+23		1.0	+19	
	Settlement, support 4 (10 mm)	+30		1.2	+36		1.0	+30	
					2136			1769	

Commentary to calculation sheet

Secondary shrinkage moments are hogging, secondary differential temperature moments are sagging.

In the midspan region, combination 3, with differential temperature, is slightly more onerous than combination 1.

The inclusion of t_f in the expression for λ_{LT} (via the ratio λ_F) is on account of the torsional stiffness of the beam. It is therefore not necessary to 'smear' the top flange of the beam across the width of the slab. The whole empirical expression for λ_{LT} generally underestimates the torsional stiffness of a composite section. It is acceptable to use the mean value of t_f as calculated opposite.

Note that the evaluation neglects the restraint to lateral distortional buckling (effectively the restraint offered by continuous U-frame restraint provided by the slab and the beam web) and is thus conservative. [The note to Clause 3/9.7.2 allows I_t to be based on 1/4 of the value for the 11.4 m wide deck but this makes very little difference because the value of (r_y/v) does not change significantly.]

Slenderness, λ_{LT}, Clause 3/9.7.

Limiting moment of resistance, M_R, Clause 3/9.8.

Bending resistance, M_D, Clause 3/9.9.1.2.

The Steel Construction Institute

Silwood Park, Ascot, Berks SL5 7QN
Telephone: (01344) 623345
Fax: (01344) 622944

CALCULATION SHEET

8. BENDING RESISTANCE

Plastic moment resistances of cross sections are:

Hogging: $M_{pe} = 12.09 \times 10^6 \times 345 = 4170$ kNm (Z_{pe} from Sheet 7)

Sagging: $M_{pe} = 15.11 \times 10^6 \times 345 = 5210$ kNm

Design resistances, if no reduction for LTB

Hogging: $4170 / (1.05 \times 1.1) = 3610$ kNm

Sagging: $5210 / (1.05 \times 1.1) = 4510$ kNm

Maximum total ULS moment at support $= -2665 - 308 \times 1.2$ (shrink) $= -3035$ kNm

Maximum total ULS moment at midspan $= 2136$ kNm (combination 1)

Ditto (combination 3) $= 644 + 1103 \times 1.1 + 59 + 253 \times 1.0$ (DT) $= 2169$ kNm

Resistance OK in midspan at ULS, but will need to check for reduced ULS resistance at supports, allowing for LTB. Because compact resistance utilised, check both regions at SLS

<u>ULS design resistance at supports</u>

Section properties for evaluation of slenderness

Consider reinforcement smeared over width of 2650: $(2924 + 5198) / 2650 = 3.06$ mm thick

For transverse bending

I_{yy} (bottom) $= 300^3 \times 25 / 12 = 5.63 \times 10^7$ mm^4

I_{yy} (top) $= 2650^3 \times 3.06 / 12 + 5.63 \times 10^7 = 4.80 \times 10^9$ mm^4

Total $I_{yy} = 4.86 \times 10^9$

$r_y = \sqrt{4.86 \times 10^9 / 37723} = 359$ mm (A from sheet 6)

For fully effective bracing at 3300 mm from support

Mean thickness of 25 and 25 + 3.06 is 26.53 mm

$\lambda = 3300 / 359 \times 26.53 / 1015 = 0.240$

$i = 5.63 \times 10^7 / 4.86 \times 10^9 = 0.01159$

$\psi_i = 2i - 1 = 0.02318 - 1 = -0.9768$

Hence $v = \left[(4 \times 0.01159 \times 0.9884 + 0.05 \times 0.240^2 + 0.9768^2)^{0.5}\ 0.9766\right]^{-0.5}$

$= 6.37$

Taking $M_B / M_A = 0$ and $M_A / M_M = -\infty$ (conservative)

$\eta = 0.77$

$k_4 = 1.0$

$\lambda_{LT} = \ell_e / r_y\, k_4\, \eta\, v = 3300 / 359 \times 1.0 \times 0.77 \times 6.37 = 45$

For Figure 11, $\lambda_{LT}\sqrt{345/355} = 45 \times 0.986 = 44$

For this value, Figure 11 gives $M_R / M_{ult} = 0.86$

Hence $M_R = 0.86 \times 4170 = 3586$ kNm

And $M_D = 3586 / (1.05 \times 1.1) = 3105$ kNm <u>Just OK</u>

Commentary to calculation sheet

Because the design resistance at ULS is based on compact sections, Clause 3/9.9.8 requires that checks are made at SLS.

Effective breadth of section allowing for shear lag is determined in accordance with Clauses 3/8.2 and 5/5.2.3.2.

Section properties for sections with shear lag were calculated in a similar manner to those given on Sheet 6. The calculations are not given here.

At SLS the beams are treated as non-compact (Clause 3/9.9.8) and for multi-stage construction stresses are determined by summation of stress distributions at the various stages of construction (see Clause 3/9.9.5.4).

Limiting stress for reinforcement at SLS is given by Clause 4/4.1.1.3 and 4/Table 2.

The effects of differential temperature at SLS include the partial factor of 0.8.

Limiting stress for concrete at SLS is given by Clause 4/4.1.1.3 and 4/Table 2.

Check support region at SLS

Effects of shear lag:
Breadth/span ratio = 1325 / 20300 = 0.065
At support ψ = 0.53 (3/Table 5, by interpolation)
For a cracked composite section $\psi' = 0.53 + (1 - 0.53) / 3$ = 0.69
At midspan ψ = 0.93 (3/Table 5, by interpolation)

	Moment	Top flange		Bottom flange		Top bars		Bottom bars	
		Z	Stress	Z	Stress	Z	Stress	Z	Stress
Load on steel section	-757	7.15	-106	-7.15	106				
Load on long term	-427	12.76	-33	-8.45	51	8.26	-52	10.33	-41
Load on short term	-1029	12.76	-81	-8.45	122	8.26	-125	10.33	-100
Shrinkage	-308	15.49	(-20)	-8.81	35	9.70	(-32)	12.31	(-25)
Shrinkage primary stresses			(28)		-9		(36)		(31)
Totals			-220		305		-177		-141

Limiting stresses: *Structural steel* $345\ N/mm^2$
Reinforcement $460 \times 0.75 = 345\ N/mm^2$

Check midspan region at SLS

Shrinkage is not adverse, but differential temperature is; check for combinations 1 and 3
Combination 3 is slightly more severe: results are as follows:

Use properties for section with shear lag

	Moment	Bottom flange		Top of slab	
		Z	Stress	Z	Stress
Load on steel section	367	-7.15	-51		
Load on long term	193	-10.88	-18	309.0	0.6
Load on short term	1209	-11.55	-105	239.2	5.1
Secondary positive DT	202	-11.55	-18	239.2	0.8
Primary positive DT			6		1.8
Totals			-18.6		8.3

Limiting stresses: *Structural steel* $345\ N/mm^2$
Concrete $40 \times 0.5 = 20\ N/mm^2$

Commentary to calculation sheet

The reinforcement modulus (200 N/mm²) is used here, because the stresses in the bars have been calculated using section moduli expressed in terms of values appropriate to reinforcement.

Crack width limitations, Clause 4/5.8.8.2 and 4/4.1.1.1.

Where staged construction is used, and the concrete is placed over the supports after that in span regions, M_g is usually smaller than M_q and hence E_m is simply equal to E_1.

Crack widths need only be calculated at the level of cover specified in Part 4, not the increased cover specified in BD 57 (see BA 57 and 4/5.8.8.2).

The stresses used for determining crack width were based on load effects for 30 units of HB although only 25 units needed to be applied when the actual bridge was designed. See further comment in SCI P289.

The crack width is just satisfactory. A smaller value would be achieved if larger bars were used (T16 in place of T12) or if the spacing were reduced.

The simplified procedure of Clause 10/8.2 is used.

Maximum and minimum values of bending moment due to the fatigue vehicle were calculated at 148 and −13 kN respectively for this example.

Section moduli are for cracked section with shear lag.

Fatigue classification of details (Part 10, Table 17).

The stiffener to flange weld is class F if the toe is at least 10 mm clear of the edge of the flange, otherwise it would be class G (see Table 17b).

Welding the upper bearing plate to the flange with an all-round weld would produce a class G detail. If this is not acceptable, the bearing plate could be bolted to the flange instead.

The Steel Construction Institute

Silwood Park, Ascot, Berks SL5 7QN
Telephone: (01344) 623345
Fax: (01344) 622944

CALCULATION SHEET

Job No:	BCR825	Sheet 18 of 32	Rev B
Job Title	Design of composite bridges – Worked example no. 1		
Subject	Bending resistance		
Client	SCI	Made by DCI	Date Sep 2000
		Checked by NK	Date Dec 2000

Check on concrete crack width

Reinforcement stresses:

Top = $177\ N/mm^2$

Bottom = $141\ N/mm^2$ *Strains:* $\epsilon = \sigma_1 / E$

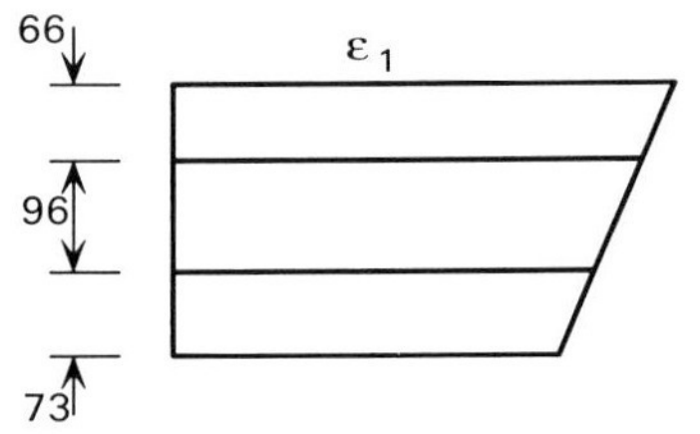

$$\epsilon = \frac{177}{200 \times 10^3} = 0.885 \times 10^{-3}$$

$$\epsilon = \frac{141}{200 \times 10^3} = 0.705 \times 10^{-3}$$

$$\epsilon_1 = \left(0.855 + 0.180 \times \frac{56}{96}\right) 10^{-3} = 0.99 \times 10^{-3}$$

Since $\frac{M_q}{M_g} > 1.0$ $\epsilon_m = \epsilon_1$

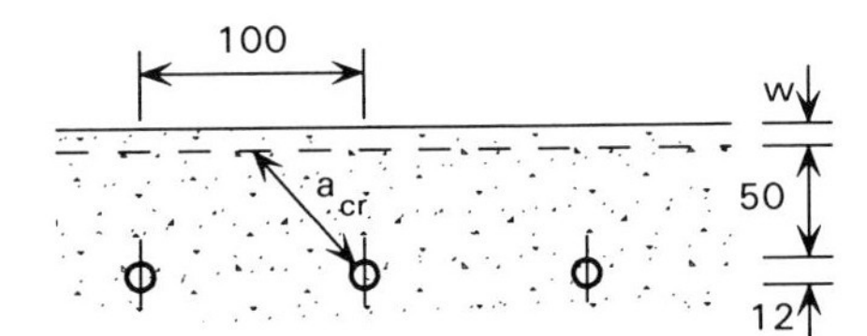

$w = 3\ a_{cr}\ .\ \epsilon_m$

$a_{cr} = (50^2 + 56^2)^{½} - 6 = 69.1\ mm$

$w = 3 \times 69.1 \times 0.99 \times 10^{-3} = 0.21\ mm < 0.25\ mm$ *Satisfactory*

Fatigue

Consider the lower flange of the beam at intermediate support
Beam section is always in hogging

Moment range due to fatigue vehicle $= 148 - (-13) = 161\ kNm$

Stress range ~ *upper surface of flange* $\frac{161 \times 10^6}{8.93 \times 10^6} = 18.0\ N/mm^2$

~ *lower surface of flange* $\frac{161 \times 10^6}{8.45 \times 10^6} = 19.1\ N/mm^2$

For class F detail, with $L = 20.3\ m$, $\sigma_H = 30\ N/mm_2$ *(Figure 8d)*
For class G detail, with $L = 20.3\ m$, $\sigma_H = 22\ N/mm_2$ *(Figure 8d)*

Hence stiffener to flange weld is satisfactory (class F detail)
Bearing plate can be welded to underside of bottom flange (class G detail)

Commentary to calculation sheet

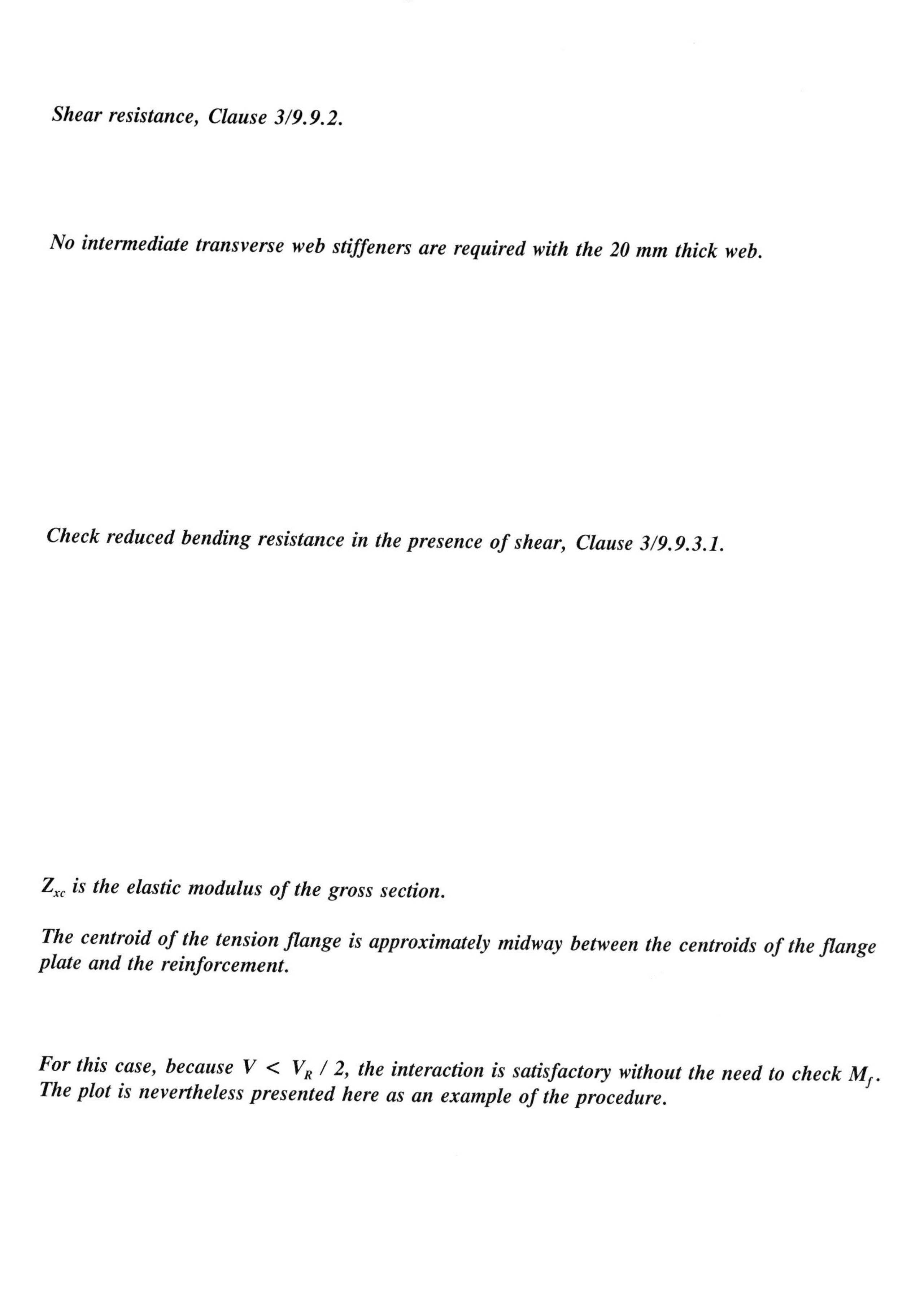

Shear resistance, Clause 3/9.9.2.

No intermediate transverse web stiffeners are required with the 20 mm thick web.

Check reduced bending resistance in the presence of shear, Clause 3/9.9.3.1.

Z_{xc} *is the elastic modulus of the gross section.*

The centroid of the tension flange is approximately midway between the centroids of the flange plate and the reinforcement.

For this case, because $V < V_R / 2$*, the interaction is satisfactory without the need to check* M_f*. The plot is nevertheless presented here as an example of the procedure.*

The Steel Construction Institute

Silwood Park, Ascot, Berks SL5 7QN
Telephone: (01344) 623345
Fax: (01344) 622944

CALCULATION SHEET

9. SHEAR RESISTANCE AND SHEAR/MOMENT INTERACTION

Shear resistance

$$\lambda = \frac{d_{we}}{t_w}\left(\frac{\sigma_{yw}}{355}\right)^{1/2} = \frac{730}{20}\left(\frac{345}{355}\right)^{1/2} = 36.0$$

Hence $\dfrac{\tau_\ell}{\tau_y} = 1.0$ (irrespective of m_{fw} and ϕ)

$$\tau_\ell = \frac{345}{\sqrt{3}} = 199\ \text{N/mm}^2$$

$$V_D = \left[\frac{t_w\,(d_w - h_h)}{\gamma_m\,\gamma_{f3}}\right]\tau_\ell = \frac{20 \times 730}{1.05 \times 1.1} \times 199 = 2.52 \times 10^6\ \text{N}$$

or $V_D = 2520$ kN $\quad (= V_R$, for $m_{fw} = 0)$

Shear/moment interaction

Take above values for M_D, V_D, V_R and plot interaction envelope

To calculate M_f : σ_f = lesser of $M_R / Z_{xc} = 3568 \times 10^6 / 8.82 \times 10^6 = 405\ \text{N/mm}^2$
and $\sigma_y = 345\ \text{N/mm}^2$

Hence $$M_f = \frac{345 \times (300 \times 25) \times 820}{1.05 \times 1.1} = 1837\ \text{kNm}$$

Load effects at ULS:

worst moment: $M = 3035,\quad V = 799$
worst shear: $M = 2821,\quad V = 1063$

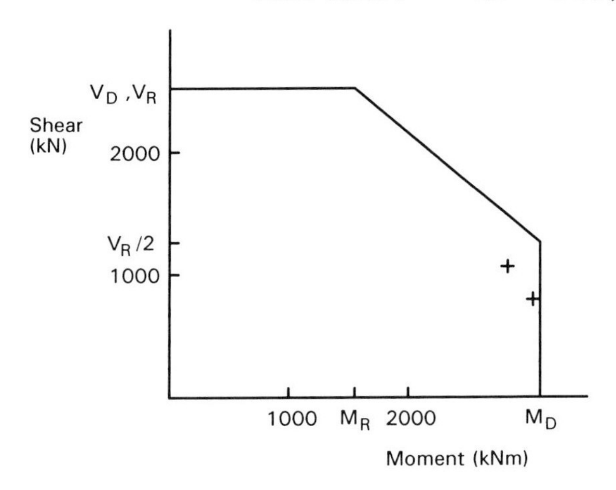

∴ *Combined shear and moment are satisfactory in both cases*

Commentary to calculation sheet

The shear connection must be checked at ULS, because construction occurs in stages (see 5/6.1.3). Shear lag is neglected at ULS (Clause 5/6.3.1).

The determination of shear flow is based on elastic uncracked section properties (see 5/5.3.1). At a support this generally gives a higher shear flow than would exist once full plastic moment had developed. This can be verified in any particular case by considering the length over which the applied moment would increase from (factored) elastic resistance to plastic resistance (of the cracked section) and the change in force in the bars that would occur over that length.

Effective breadth for calculation of shear flow is that for the quarter-span position, as given by Clause 5/5.3.1.

10. SHEAR CONNECTION

Ultimate limit state

Determine value of $\frac{A\,\bar{y}}{I}$ *for calculation of shear flow:*

Inertia of gross section = 9.02×10^9 mm^4 (Sheet 6)
Height of NA = 776.2 mm

At steel/concrete interface

$$\frac{A\,\bar{y}}{I} = \frac{401 \times 235 \times 121.3}{9.02 \times 10^9}$$

$$= 1.267 \times 10^{-3}\ \text{mm}^{-1}$$

Serviceability limit state

Determine value of $\frac{A\,\bar{y}}{I}$ *for effective sections*

Effective width of slab at 1/4 span

Spans 2 and 3 $\quad \frac{b}{L} = \frac{1325}{20300} = 0.065$, hence $\psi = 0.80$

$b_e = 0.80 \times 1325 = 1060$ mm (160 mm steel units)

Spans 1 and 4 $\quad \frac{b}{L} = \frac{1325}{14000} = 0.095$, hence $\psi = 0.764$ (3/4 span)

$\psi = 1.0$ (1/4 span)

For $\psi = 0.80$, $\quad I = 8.19 \times 10^9$ and $y_{na} = 759.7$ mm

$$\frac{A\,\bar{y}}{I} \text{ at interface} = \frac{320 \times 235 \times 137.8}{8.19 \times 10^9} = 1.265 \times 10^{-3}$$

Very little difference, so use 1.27×10^{-3} *generally*

Commentary to calculation sheet

Values of shear for the various positions in the spans were determined by the computer analyses for this example.

Values of γ_m given in BD 16/82 and the 'yellow book' are:

SLS: γ_m = 1.85 (Clause 5/5.3.2.5)
ULS: γ_m = 1.40 (Clause 5/6.3.4)

The design shear flows are factored by γ_m γ_{f3} to facilitate comparison between the requirements for static strengths at ULS, at SLS and for fatigue strength.

Effectively this compares:

(nominal loads $\times$ γ_{f3} $\times$ γ_{fL} $\times$ γ_m) with (nominal static strength)

Silwood Park, Ascot, Berks SL5 7QN
Telephone: (01344) 623345
Fax: (01344) 622944

CALCULATION SHEET

Job No: *BCR825*	Sheet *21* of *32*	Rev *B*
Job Title *Design of composite bridges – Worked example no. 1*		
Subject *Shear connection*		
Client *SCI*	Made by *DCI*	Date *Sep 2000*
	Checked by *NK*	Date *Dec 2000*

Shear loading on spans 3 and 4

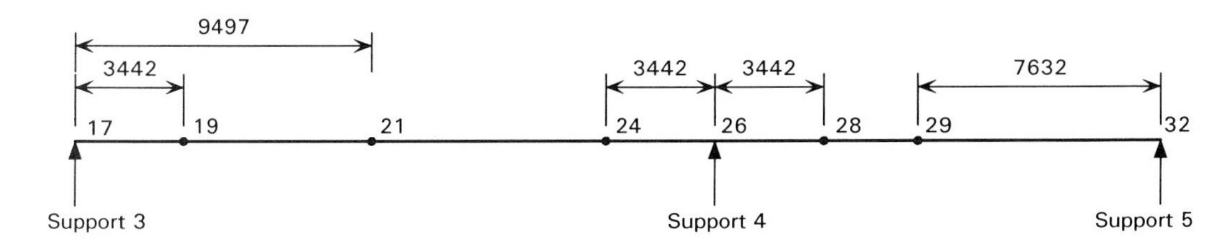

Loading \ Node	17	19	21	24	26 LHS	26 RHS	28	29	32
ULS									
Dead load	372	250	35	-226	-348	312	189	86	-185
HB+HA (max. shear cases)	673	437	-212	-400	-590	601	424	-190	-526
Footway	-	7		-9					
Differential settlement	17	17	-17	-17	-17	12	12	-12	-12
Shrinkage	-		-6	-6	-6	35	35		
Total shear (V) (kN)	1062	711	-200	-658	-961	960	660	-116	-723
Shear flow $(VAy\ /\ I)$	1349	903	254	836	1220	1219	838	147	918
Shear flow $\times\ \gamma_m\ \gamma_{f3}$ *(N/mm)*	2077	1391	391	1287	1879	1877	1291	226	1414
SLS									
Remaining concrete	0	-9	-9	-9	0	0	-2	-2	-2
SDL	92	62	9	-56	-86	77	47	21	-48
HB+HA	570	370	-179	-339	-499	508	359	-161	-445
Footway		5		-6					
Differential settlement	14	14	-14	-14	-14	10	10	-10	-10
Shrinkage	-	-	-5	-5	-5	29	29	-	-
Total shear (V) (kN)	676	442	-198	-429	-604	624	443	-152	-505
Shear flow $(VAy\ /\ I)$	859	561	251	545	767	792	563	193	641
Shear flow $\times\ \gamma_m\ \gamma_{f3}$ *(N/mm)*	1589	1038	464	1008	1419	1465	1042	357	1186

Commentary to calculation sheet

Stresses in the weld attaching shear connector are given by Clause 10/6.4, and the simplified procedure for checking fatigue of the shear connection is given in Clause 10/8.2.

Values for the range of vertical shear due to the passage of the fatigue vehicle were determined by grillage analysis.

Base lengths of the influence line (positive or negative area for maximum effect) were determined by line-beam analysis.

The equivalent shear flow determined opposite is the shear resistance when the connectors are just adequate to provide the necessary fatigue life.

Static strengths of shear connectors are given in Clause 5/5.3.2 and Table 7.

Fatigue on shear studs

Node	17	21	26	29	32
Range of vertical shear under fatigue vehicle (V) (kN)	133	98	135	85	137*
Longitudinal shear range $\left(\frac{VAy}{I}\right)$ *(N/mm)*	169	124	171	108	174
Base length of influence line (m)	40.6	10.2	34.3	7.6	14.3
Limiting stress range σ_H *(N/mm²)*	43	59	46	60	57
Equivalent static shear flow (N/mm)	1670	893	1580	765	1300

* *Including impact*

Notes:

1 *Range of vertical shear from global analysis*

2 $\frac{A_y}{I}$ *taken as* $1.27 \times 10^{-3}\ mm^{-1}$

3 σ_H *from Figure 8(d), Part 10, using values of base length of influence line*

4 *Static strength of stud* $= 425\ N/mm^2$

5 *Equivalent static shear flow = longitudinal shear range × 425 / limiting stress range*

Commentary to calculation sheet

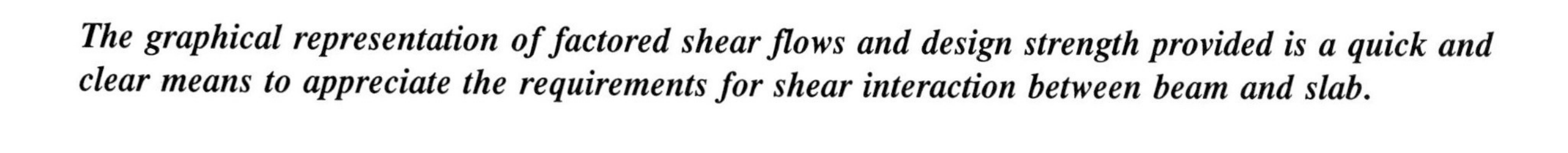

The graphical representation of factored shear flows and design strength provided is a quick and clear means to appreciate the requirements for shear interaction between beam and slab.

The line shown for 'design provision' is determined on the basis of the nominal static strength for 19 mm studs, 100 mm high, as given in Part 5, Table 7. All partial factors have been applied to the loadings (i.e. the plotted lines); the static strength of the studs is the unfactored nominal static strength, assuming that the studs are attached in pairs at the intervals indicated.

Transverse reinforcement must be checked at ULS for the calculated values of shear flow (Clause 5/6.3.3). The check is not shown here.

Silwood Park, Ascot, Berks SL5 7QN
Telephone: (01344) 623345
Fax: (01344) 622944

CALCULATION SHEET

Job No: *BCR825*		Sheet *23* of *32*	Rev *B*
Job Title *Design of composite bridges – Worked example no. 1*			
Subject *Shear connection*			
Client *SCI*	Made by *DCI*	Date *Sep 2000*	
	Checked by *NK*	Date *Dec 2000*	

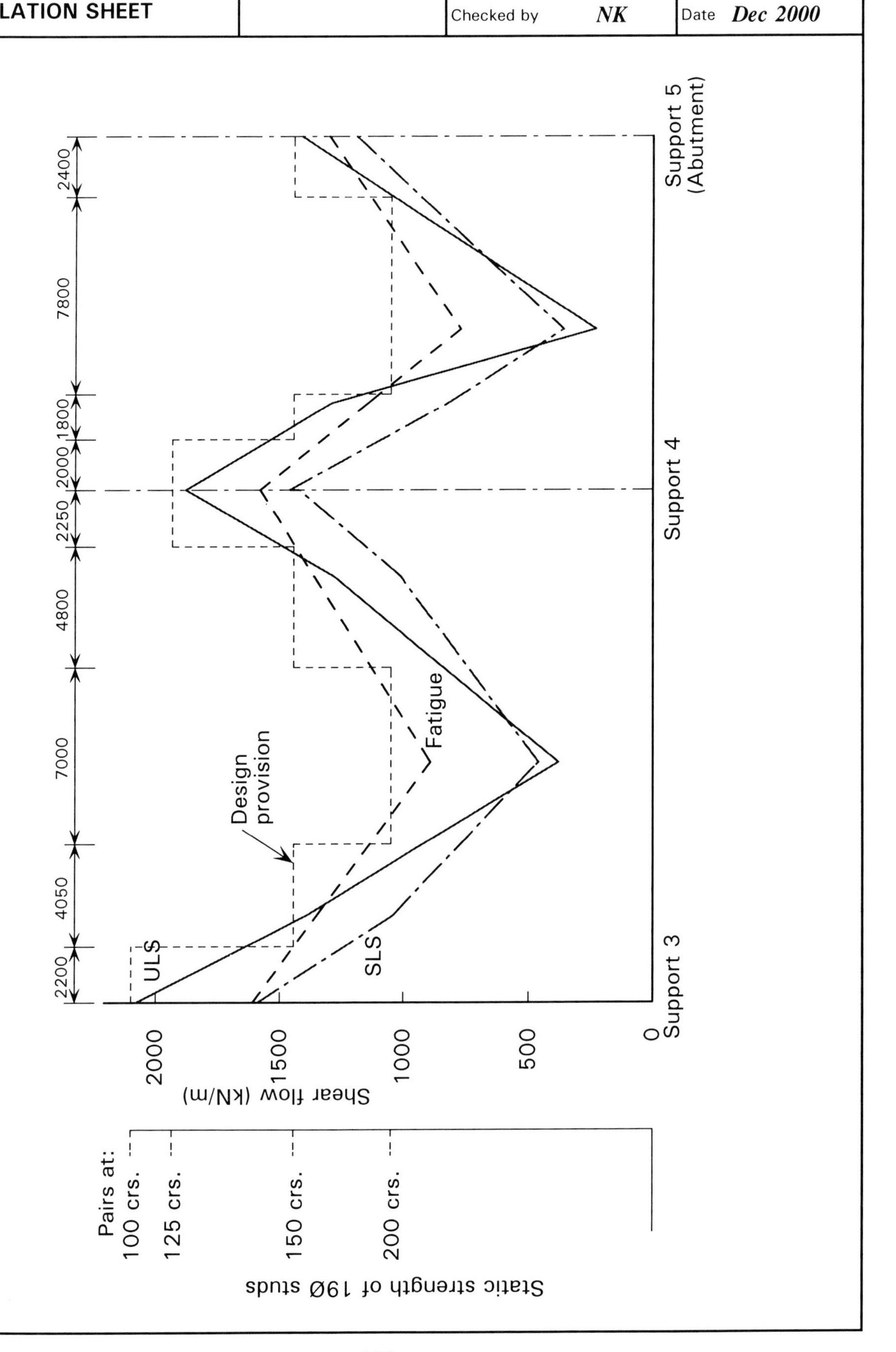

Commentary to calculation sheet

Load effects at the splice were determined from results of line-beam and grillage analyses for this example.

It may be noted that the ULS moment, determined from an analysis with all the loading applied to a composite, complete, structure, is about 40% of the design moment at support number 3.

The splice only needs to be designed to carry the forces at that position, not to transmit the full strength of the beam.

11. SPLICES

Splices are located 2.6 m to one side of each internal support

Consider splice near support No. 3, within span 3

Loading effects

Loading	Nominal		ULS			SLS		
	BM	Shear	γ_{fL}	BM	Shear	γ_{fL}	BM	Shear
Total effects on bare steel beam				-328	176		-288	154
Total effect on composite section				-105	104		-73	71
HB 30 (in span 2)	-365	26	1.3	-475	34	1.1	-402	29
HA (lane 2, span 2)	-175	15	1.3	-228	20	1.1	-193	17
FW (next to lane 2, span 2)	-11	1	1.5	-16	2	1.0	-11	1
				-1152	336		-967	272
Settlement (supports 2 and 4)	-102	14	1.2	-122	17		-102	14
				-1274	353		-1069	286

Ultimate limit state

Consider DL + SL effects all on composite section (compatible with support and midspan design)
Use elastic section properties (hinges not formed)
Gross section

Z_{bot} = 8.81×10^6 mm^3 (derived from properties on Sheet 6)
$Z_{top\ flge}$ = 15.49×10^6 mm^3

$$\sigma_b = \frac{1274 \times 10^6}{8.81 \times 10^6} = 145 \text{ N/mm}^2$$

$$\sigma_t = -\frac{1274 \times 10^6}{15.49 \times 10^6} = -82 \text{ N/mm}^2$$

At the top and bottom of web, stresses are –75 and 138 N/mm^2

Serviceability limit state

$$\sigma_b = \frac{1069 \times 10^6}{8.81 \times 10^6} = 121 \text{ N/mm}^2$$

$$\sigma_t = -\frac{1069 \times 10^6}{15.49 \times 10^6} = -69 \text{ N/mm}^2$$

Commentary to calculation sheet

The diagrams opposite show the configuration of the splices. The splices were designed on the basis of no slip at SLS and were checked for bearing/shear at ULS. For typical calculation for splice design, see Worked example no. 2.

For a quick comparison, the load in 100 mm of web at 100 N/mm^2 (approximate average stress at bottom of web at SLS) = 200 kN. Friction capacity of a M30 HSFG bolt in double shear = 215 kN at SLS. Load in flange = 1090 kN at ULS, 910 kN at SLS.

Silwood Park, Ascot, Berks SL5 7QN
Telephone: (01344) 623345
Fax: (01344) 622944

CALCULATION SHEET

Job No: *BCR825*	Sheet *25* of *32*	Rev *B*
Job Title *Design of composite bridges – Worked example no. 1*		
Subject *Splices*		
Client *SCI*	Made by *DCI*	Date *Sep 2000*
	Checked by *NK*	Date *Dec 2000*

Web splice

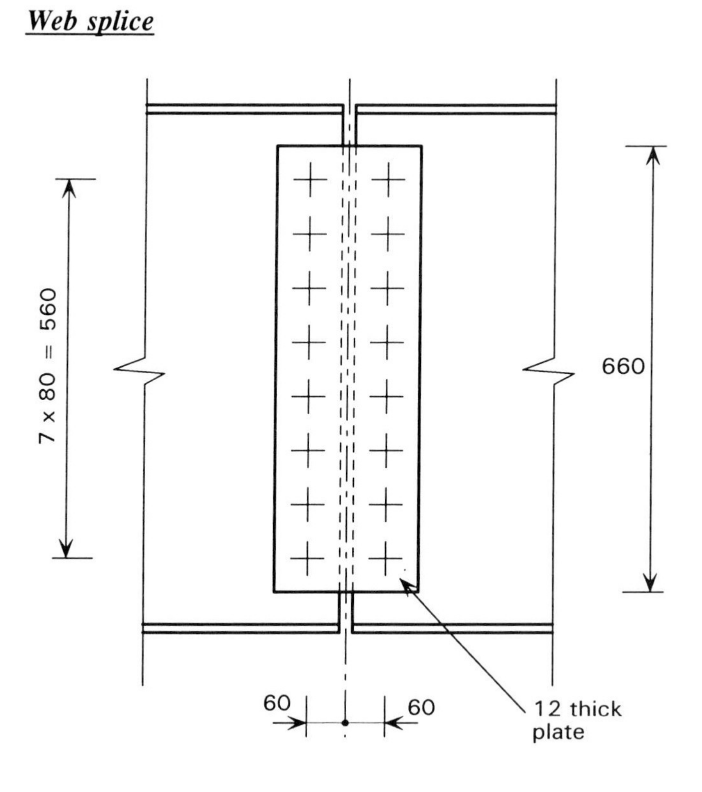

Web covers:
220 × 12 × 660 mm

8 No. M30 HSFG bolts

Flange splice

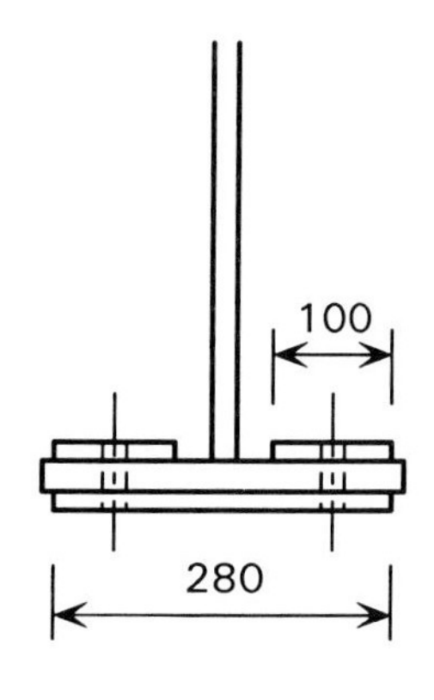

Top covers: *100 × 15 × 560 mm*
Bottom covers: *280 × 15 × 560 mm*

3 Rows of 2 No. M30 HSFG bolts each side of splice

Commentary to calculation sheet

The bracing system at one internal support is shown opposite. Only one frame is considered here in detail. A similar check must be made on all frames, though the designer should be able to identify where comparison with other frames will permit checking by inspection only.

The complete bracing system must provide sufficient restraint to the main beams during construction, particularly under the weight of the wet concrete. Separate calculations show that the channel bracing, which acts as torsional restraint to the main beams, is insufficient to reduce the LTB effective length to a low enough value that the design moment resistance is adequate for the wet concrete condition. A more effective system would be plan bracing, possibly as a series of diagonals across all the beams on top of the top flange. The design of that bracing is not covered in this example (see Worked example no. 2 for calculations for torsional restraint to main beams).

Small deflections of the outer beam (of the order of 5 mm) will lead to a twisting about the longitudinal axis. Pot-type bearings will permit small rotations without contributing to any eccentricity of the centre of bearing pressure.

Silwood Park, Ascot, Berks SL5 7QN
Telephone: (01344) 623345
Fax: (01344) 622944

CALCULATION SHEET

Job No:	BCR825	Sheet	26 of 32	Rev	B
Job Title	Design of composite bridges – Worked example no. 1				
Subject	Bracing				
Client	SCI	Made by	DCI	Date	Sep 2000
		Checked by	NK	Date	Dec 2000

12. BRACING

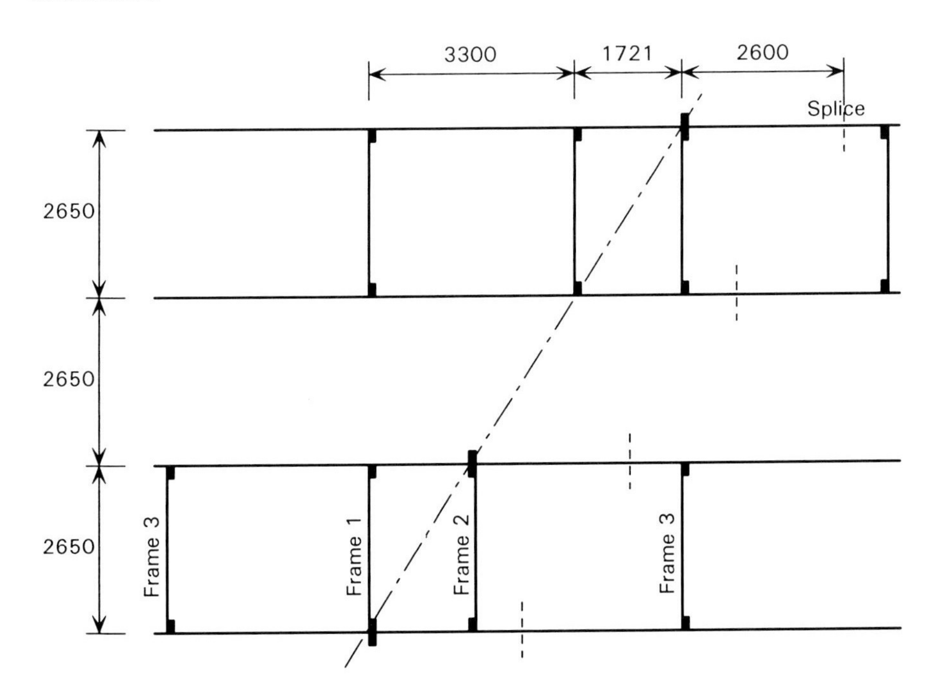

Plan arrangement of bracing at intermediate supports

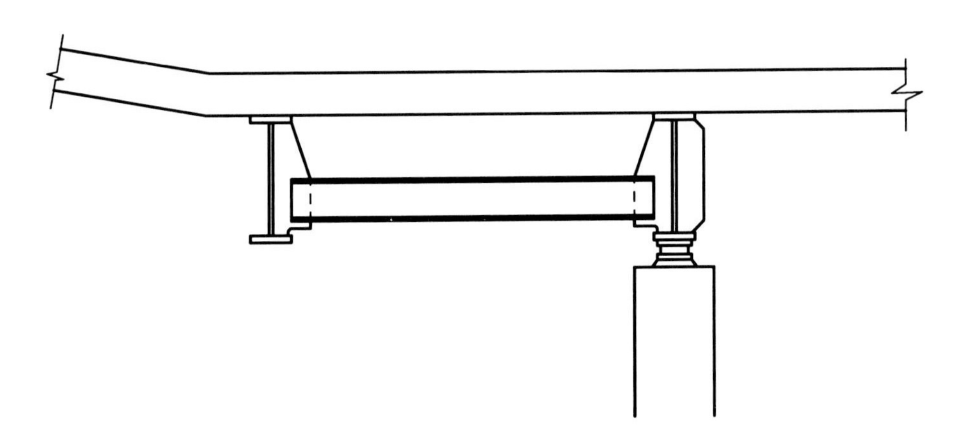

Elevation on frame 2

Bearings selected are pot type sliding bearings; one bearing at each support is guided, therefore it will transmit lateral loads. Rotation capacity of bearing (0.01 rad) will accommodate differential deflection of outer girders.

Commentary to calculation sheet

Setting eccentricities are not given for pot-type bearings in Clause 3/9.14.3.3, but a value of 10 mm in either direction seems reasonable here.

The cross member provides torsional restraint to the main beams and is therefore required to be designed to resist a couple, determined as a pair of equal and opposite lateral forces in each flange.

If the cross member were restraining two main beams at bearing supports, one pair of forces, F_S*, would need to be considered at each end of the cross member (Clause 3/9.12.5.2).*

The Steel Construction Institute

Silwood Park, Ascot, Berks SL5 7QN
Telephone: (01344) 623345
Fax: (01344) 622944

CALCULATION SHEET

Job No: *BCR825*	Sheet *27* of *32*	Rev *B*
Job Title *Design of composite bridges – Worked example no. 1*		
Subject *Bracing*		
Client *SCI*	Made by *DCI*	Date *Sep 2000*
	Checked by *NK*	Date *Dec 2000*

Bracing frame no. 2 (at support no. 3)

Load effects

(i) Vertical reactions

	Load combination 1				Load combinations 2 and 3			
	Loading for max. R		Loading for max. M		Loading for max. R		Loading for max. M	
	R	Coex. M	Coex. R	M	R	Coex. M	Coex. R	M
DL+SL	720	1086	720	1086	720	1086	720	1086
HB+HA	857	821	756	1192	725	695	640	1009
F/W	17	23	-6	30	14	19	-5	25
Settlement	17	83	17	83	17	83	17	83
Shrinkage	-	-	-11	266	-	-	-11	266
	1611	2013	1476	2657	1476	1883	1361	2469

(ii) Bending due to eccentricity of bearing

Setting eccentricity $\pm$ *10 mm (transverse and longitudinal)*

Thermal movement from datum (20 °C) to U_e *(−14.5 °C)*
$= 34.5 \times 12 \times 10^{-6} \times 34300 = 14$ *mm*

At ULS $\gamma_{fl} = 1.0$, *so* $e =$ *24 mm longitudinally and*
10 mm transversely

(iii) Bending due to restraint forces

Bracing system must be designed to restrain torsion of the beam against a couple, equal and opposite forces at flange level

F_S *is the sum of up to four components (3/9.12.5.2)*

$$F_S = F_{S1} + F_{S2} + F_{S3} + F_{S4}$$

Commentary to calculation sheet

F_{S1} is the force due to the bow of the compression flange (3/9.12.5.2.1).

σ_{fc} is taken as factored yield, rather than working through the load effects to determine the actual (lower) value of the stress.

F_{S2} is the force due to the non-verticality of the web (3/9.12.5.2.2).
This does not include out-of-vertical due to twist at skew supports, which is given by F_{S4}.

Calculate the four components of F_S according to 3/9.12.5.2

$$F_{S1} = \frac{0.006\,M}{d_f[1 - (\sigma_{fc}/\sigma_{ci})^2]}$$

Take $\sigma_{fc} = \sigma_y / \gamma_m \gamma_{f3} = 299\ N/mm^2$

$$S = Z_{pe} / Z_{xc} = 12.09 \times 10^6 / 8.81 \times 10^6 = 1.37$$

$M = 3035\ kNm$ *(Sheet 16)*

$\lambda_{LT} = 45$ *(Sheet 16)*

$$\sigma_{ci} = \frac{\pi^2 ES}{\lambda_{LT}^2} = \frac{\pi^2 \times 205000 \times 1.37}{45^2} = 1370\ N/mm^2$$

$$F_{S1} = \frac{0.006 \times 3035 \times 10^6}{1015[1 - (299/1370)^2]} = 18.8\ kN$$

$$F_{S2} = \frac{\beta(\Delta_{e1} - \Delta_{e2})}{(\sigma_{ci} - \sigma_{fc})\sum\delta}$$

$\Delta_{e1} = \Delta_{e2} = D/200 = 5\ mm \qquad \beta = 2.0$ *(internal support)*

For δ_i, consider the restraint provided by the channel. For a moment M on the end of a beam of length L, second moment of area I, the rotation is given by:

$$\theta = ML/3EI$$

So, for unit forces a distance d apart, the relative displacement is

$$\delta = d^2L/3EI$$

Hence, for a channel of length 2650 and $I = 44.5 \times 10^6\ mm^4$

$$\delta_i = \frac{2650 \times 800^2}{3 \times 205000 \times 44.5 \times 10^6} = 6.2 \times 10^{-5}\ mm$$

To determine σ_{ci} for F_{S2}, calculate ℓ_e according to 3/9.6.2(b)

$\Sigma L = 20.3 + 20.3 = 40.6\ m$

$I_c = 56.25 \times 10^6\ mm^4$

Take $k_1 k_2 = 1.0$

Commentary to calculation sheet

F_{S3} is the force due to the eccentricity of the bearing reaction resulting from the initial out-of-verticality of the support. The out-of-vertical is taken as D/200, assuming that the structure will comply with BS 5400–6, which specifies a tolerance of D/300 (Clause 3/9.12.5.2.3).

It is possible that other load cases, for example for maximum longitudinal rotation of the beam, with coexistent reaction, would give rise to a greater value of F_{S3} and F_{S4}. A conservative view would be to consider all the maximum values as coexistent, although that has not been done here.

F_{S4} is the force due to the twisting of the beam caused by changes in the longitudinal slope of a beam on a skew support. In this case, paragraphs 3 and 4 of 3/9.12.5.2.4 apply.

$$\ell_E = \frac{40.6 \times 10^3}{\sqrt{1 + 2\dfrac{\left(40.6 \times 10^3\right)^3}{\pi^4 \times 205000 \times 56.25 \times 10^6 \times 9.3 \times 10^{-5}}}} = 1134 \text{ mm}$$

$\lambda_{LT} = 1134 / 359 \times 1.0 \times 6.37$ ($k_4\eta = 1.0$, v from Sheet 16)

$= 20.1$

$$\sigma_{ci} = \frac{\pi^2 ES}{\lambda_{LT}^2} = \frac{\pi^2 \times 205000 \times 1.37}{20.1^2} = 6860 \text{ N/mm}^2$$

$$F_{S2} = \frac{2.0 \times 10 \times 299}{(6860 - 299) \times 18.6 \times 10^{-5}} = 4900 \text{ N} = 4.9 \text{ kN}$$

$F_{S3} = Rd_L (\Delta/D + \theta_L \tan \alpha) / D$

$R = 1611$ kN (Sheet 27)

$d_L = 780$ mm

$\Delta = 1015 / 200 = 5.08$ mm

$\theta_L = 0.001$ rad (rotation at support for this loadcase)

$\alpha = 33°$

$F_{S3} = 1611 \times 10^3 \times 780 \times (1/200 + 0.001 \tan 33) / 1015$

$= 7.0$ kN

$$F_{S4} = \frac{D\theta_{LA} \tan\alpha}{\lambda\, \delta_R'}$$

$\lambda = 1 / (4\delta_R' EI_F)^{0.25}$

$\delta_R' = \delta_R = d_1^3 / 3EI_1 + uBd_2^2 / EI_2$

$d_1 = 768$ mm $d_2 = 855$ mm

$I_1 = t_w^3 / 12 = 667 \text{ mm}^4/\text{mm}$

$I_2 = (t_{slab})^3 / 12 \times E_C / E_S = 163900 \text{ mm}^4/\text{mm}$

$$\text{Hence } \delta_R = \frac{768^3}{3 \times 205000 \times 667} + \frac{0.5 \times 2650 \times 885^2}{205000 \times 163900} \text{ mm}^2/\text{N}$$

$= 1.14 \text{ mm}^2 / \text{N}$ (i.e. deflection for 1N per 1 mm length of web)

Commentary to calculation sheet

To determine the value of F_R (Clause 3/9.12.2) for the intermediate restraint to the outer girder that connects to the inner beam at the support, it is assumed that half the design moment capacity is mobilised in the outer beam at the same time as full capacity is mobilised in the inner beam.

$\lambda = 1 / (4 \times 1.14 \times 205000 \times 56.25 \times 10^6)^{0.25}$

$= 0.00037 \text{ mm}^{-1}$

Hence $F_{S4} = \dfrac{1015 \times 0.001 \times 0.649}{0.00037 \times 1.14} = 1560 \text{ N} = 1.56 \text{ kN}$

This force arises from each side of the support, so total $F_{S4} = 3.1 \text{ kN}$

Total restraint force

$F_S = 18.8 + 4.9 + 7.0 + 3.1 = 33.8 \text{ kN}$

Lateral force due to wind

Wind load on structure with live load (combination 2)

$= 3.88 \text{ kN/m} \times \gamma_{fL}$ (Sheet 14)

Load on central support of continuous beam is approximately 1.14 wL

$= 3.88 \times 1.1 \times 20.3 \times 1.14$

$= 98 \text{ kN}$

Lateral restraint force for intermediate restraint to outer girder (at S2)

Force = flange force / 80 — 3/9.12.2

$= \frac{1}{2} \times (300 \times 25 \times 299) / 80 = 14.0 \text{ kN}$

Commentary to calculation sheet

The sheet summarises the forces on the bracing system. The forces and moments in the various components of the bracing system can be calculated by a simple local plane frame analysis. The local analysis and the checking of the components are not included here.

Silwood Park, Ascot, Berks SL5 7QN
Telephone: (01344) 623345
Fax: (01344) 622944

CALCULATION SHEET

Summary of forces on frame no. 2

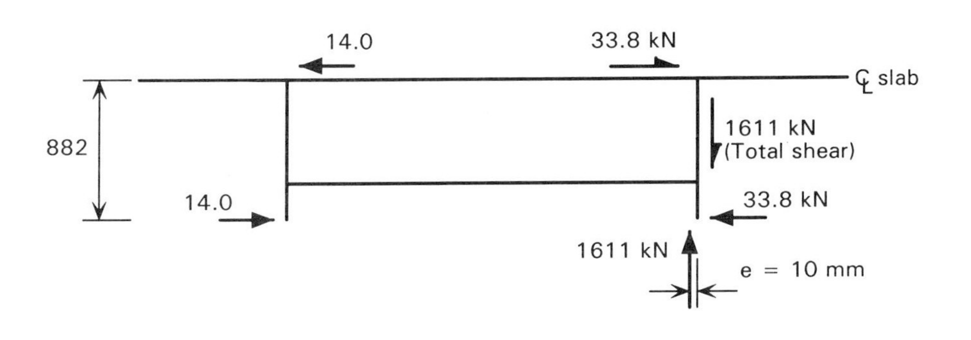

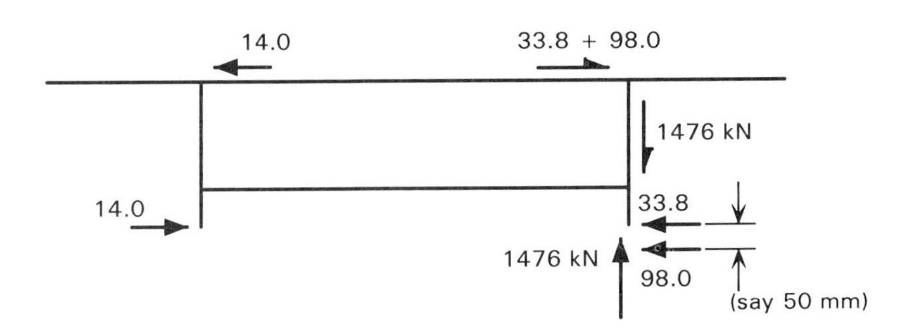

Commentary to calculation sheet

The check opposite is only for the bearing stress at the bottom of the bearing stiffener. Further checks should be made for the combination of bending and axial effects from the local analysis of the bracing frame, but those calculations are not included; the bearing stiffener is more than adequate in this case.

Just above the cross beam, the stiffener outstand on that side exceeds the usual 10t limit. If this persisted for any significant length, the effective yield of the stiffener would need to be reduced in accordance with 9.3.1 so that it did comply. But here, in view of the taper and the restraint of the cross beam, the slight infringement is accepted (alternatively, the stiffener could be cut back just above the cross beam).

On the other side, an alternative to the 45 ° snipe where the stiffener joins the flange would be a radiused notch, similar to that shown at the bottom of the stiffener that is attached to the cross beam. For further advice on shaping stiffeners, see Guidance Note 2.05.

By inspection, longitudinal bending on the effective stiffener section will be very small and does not govern the design.

The stresses in the stiffener material are sufficiently low that grade S275 material could be used in this case, but it is better not to mix material grades.

Longitudinal eccentricity = 10 mm setting error plus thermal expansion from a fixed bearing at one end (35 °C change). Transverse eccentricity = 10 mm setting error.

The Steel Construction Institute

Silwood Park, Ascot, Berks SL5 7QN
Telephone: (01344) 623345
Fax: (01344) 622944

CALCULATION SHEET

Job No:	*BCR825*	Sheet *32* of *32*	Rev *B*
Job Title	*Design of composite bridges – Worked example no. 1*		
Subject	*Bearing stiffener*		
Client *SCI*	Made by *DCI*	Date *Sep 2000*	
	Checked by *NK*	Date *Dec 2000*	

13. BEARING STIFFENER S4

Bearing stress at upper face of bottom flange

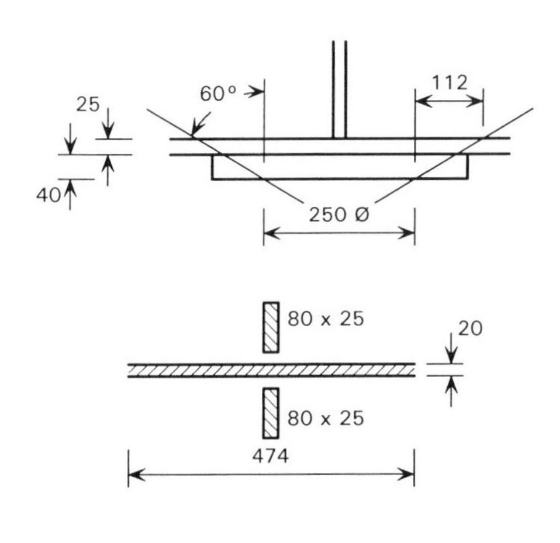

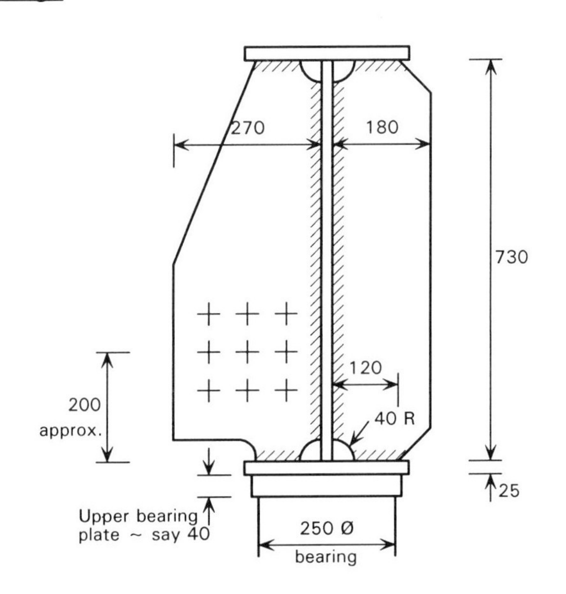

$A = 13480 \text{ mm}^2$
$I_{xx} = 3.45 \times 10^7$
$I_{yy} = 1.77 \times 10^8$

Maximum reaction = *1611 kN*

Axial stress = $1611 \times 10^3 / 13480 = 120 \text{ N/mm}^2$

Transverse bending stress = $(1611 \times 10^3) \times 10 / (3.45 \times 10^7 / 130)$

= 61 N/mm^2

Maximum stiffener stress = $120 + 61 = 181 \text{ N/mm}^2$

Longitudinal bending stress = $(1611 \times 10^3) \times 24 / (1.77 \times 10^8 / 237)$

= 52 N/mm^2

Maximum vertical stress in web

= $120 + 52 = 172 \text{ N/mm}^2$

Worked Example Number 2

Contents

Commentary to calculation sheet

The worked example is a three-span bridge carrying a wide two-lane highway over a river. Hydraulic considerations for the flood plain led to the requirement of a 90 m clearance between abutments. To keep the piers and pile caps out of the river, a 42 m centre span was selected. The side spans were made equal at 24 m, approximately 60% of the main span, and are thus a slightly lower proportion than the ideal of around 80%.

A haunched configuration was adopted for reasons of appearance and economy.

The Steel Construction Institute

Silwood Park, Ascot, Berks SL5 7QN
Telephone: (01344) 623345
Fax: (01344) 622944

CALCULATION SHEET

Job No:	*BCR825*	Sheet *1* of *43*	Rev *B*
Job Title	*Design of composite bridges – Worked example no. 2*		
Subject	*Outline drawings*		
Client	*SCI*	Made by *DCI*	Date *Sep 2000*
		Checked by *CRH*	Date *Dec 2000*

1. *OUTLINE DRAWINGS*

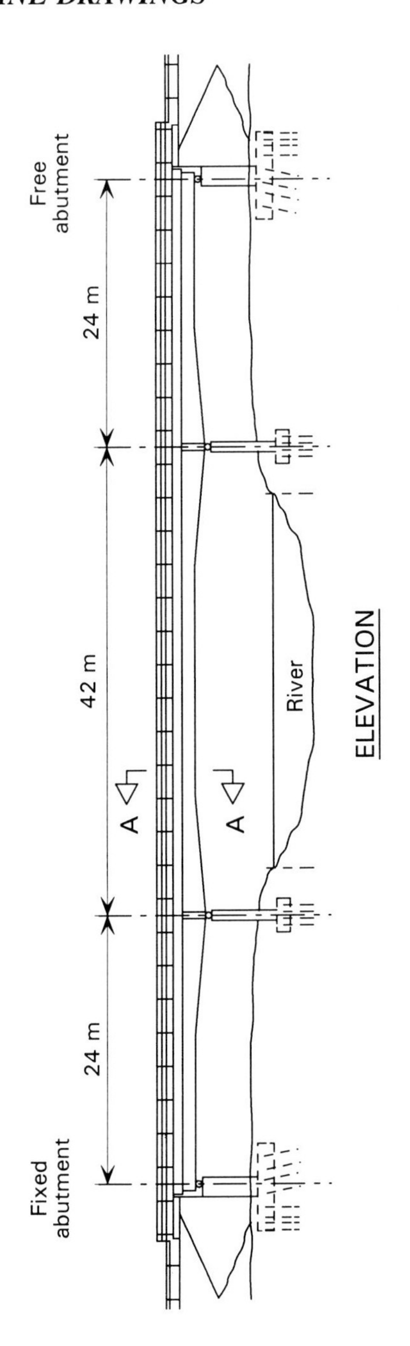

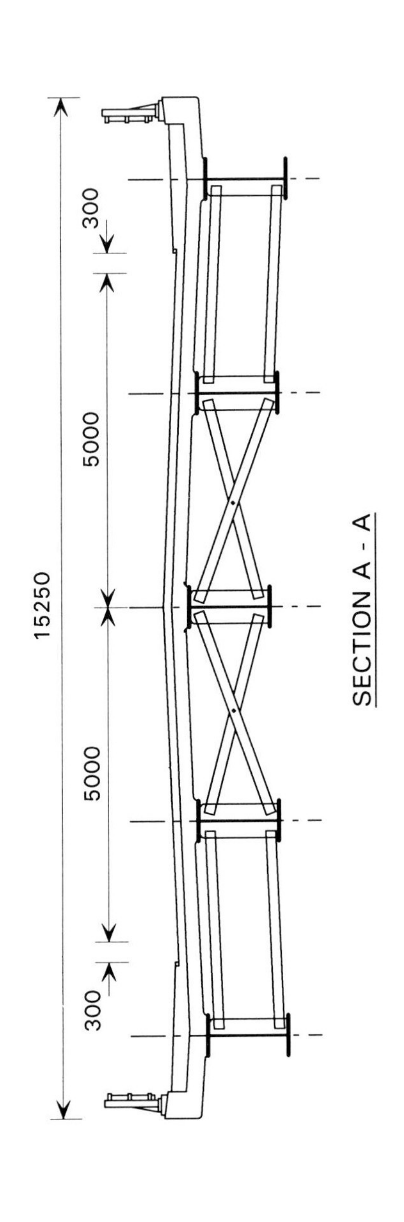

Commentary to calculation sheet

Detailed design was first carried out for a standard two-lane carriageway, and four girders were chosen to support the deck. At a later stage, the wide carriageway was specified and it was considered expedient to accommodate this simply by increasing the number of main girders to five; extra design work was thereby minimised. The worked example here starts with the final configuration of five beams and follows the design process through for one of the internal beams.

The settlement allowance shown opposite is much smaller than actually allowed in design (20 mm). The value has been reduced in this publication to ensure that the total design load effects do not exceed their design resistance to the revised code.

According to 3/6.2, the nominal yield stress to be used in design is the minimum specified yield strength for the appropriate thickness. The table below gives mechanical properties for grade S355 steel to BS EN 10025.

	Material thickness		
	t ≤ 16 mm	*16 mm > t ≤ 40 mm*	*40 mm > t ≤ 63 mm*
Yield strength	*355 N/mm²*	*345 N/mm²*	*335 N/mm²*

2. DESIGN DATA

General

Location:	*Eastern England*
Design life:	*120 years*
Carriageway:	*10 m wide, marked for two lanes with 0.3 m hard strips*
Footways:	*1.7 m – 2.0 m wide on either side (road curved in plan)*
Spans:	*24.0, 42.0, 24.0*
Skew:	*0 °*
Surfacing:	*130 mm thick (including waterproofing)*
Parapets:	*Type P2 on 475 mm wide upstand*

Loading

Dead load: *Steelwork* $77\ kN/m^3$
Concrete $26\ kN/m^3$
Surfacing $24\ kN/m^3$
Parapets $0.5\ kN/m$

Settlement: *2 mm maximum at any support*

Live load: *HA – 3 notional lanes*
HB – 45 units, in conjunction with HA
Abnormal indivisible loads (AIL)
(i) Trailer 987/14 – 300 tonnes, or
(ii) Trailer 1277/12 – 269 tonnes
[Each AIL is accompanied by two tractor units (47 t) and associated HA]
Footway loading

Temperature: *Minimum shade air temperature −18 °C*
Minimum effective bridge temperature −15 °C

Wind: *Maximum mean hourly wind speed 29 m/s*

Deck configuration

Slab: *250 mm thick, grade 40 concrete*
Girders: *5 No. girders, steel grade S355*

Design parameters

Steel: $\sigma_y = 355\ N/mm^2$ *(for thicknesses up to 16 mm)*
$E_s = 205\ kN/mm^2$
$\nu = 0.3$

Concrete: $f_{cu} = 40\ N/mm^2$
$E_{cs} = 31\ kN/mm^2$ $E_{cl} = 15.5\ kN/mm^2$
$\nu = 0.2$

Reinforcement: $f_{ry} = 460\ N/mm^2$

Coefficient of thermal expansion: $\alpha = 12 \times 10^{-6}$ *(steel and concrete)*

Commentary to calculation sheet

The end of the haunches and the splice positions were chosen to be near the point of contraflexure in the main span, and it was decided to make the haunches symmetrical about the support. See Sheet 5 for further details.

The shear in a single girder due to the HB vehicle will not be as high as 70% of the total load of that vehicle, but it is reasonable when allowing for coexistent effects from other lanes.

For initial design, shear resistance based on Clause 3/9.9.2. The actual design will need to use Clause 3/9.11, but the use of the simpler clause is adequate for initial design.

The value of 90% used for the proportion of moment carried by the flanges is arbitrary but reasonable for initial design for this configuration.

3. PRELIMINARY SIZING

Girder depth

Most economic span:depth ratio is between 20 and 25:1. However, by haunching, the depth at middle of the main span may be reduced at the expense of the depth at the piers

Midspan depth = 1200 + 275 (slab + haunch) = 1475
(Ratio span:depth = 42000:1475 = 28.5:1)

Pier depth = 2000 + 275 = 2275 (ratio = 18.5:1)

Haunches are chosen to be 9500 long

Splices are provided at the end of each haunch

Pier web sizing

For 3.2 m width of slab on each girder:

Calculate shears:					
	steel	*say 150 kN*	= *150 × 1.1*	=	*165*
	concrete	*3.2 × 0.25 × 21 × 26*	= *437 × 1.2*	=	*524*
	surfacing	*3.2 × 0.13 × 21 × 24*	= *210 × 1.75*	=	*368*
	HB	*say 70% of vehicle (= 0.7 × 1800 kN)*			
			= *1260 × 1.3*	=	*1638*
					2695 kN

Size web to take 150% of shear, say 4050 kN

Try web thickness $t_w = 16$ *mm*

Web depth $d_{we} = 1900$ *mm, hence* $d_{we} / t_w = 119$

Shear yield stress $\tau_y = \sigma_y / \sqrt{3} = 355 / \sqrt{3} = 205$ N/mm^2

Assume aspect ratio of 1.0 and unrestrained boundary for panel adjacent to pier

Then $\tau_l / \tau_y = 0.7$ *(Figure 12,* $m_{fw} = 0$*)*

Thus, shear resistance of web

$$= 0.7 \times 205 \times 1900 \times 16 \times 10^{-3} / (1.1 \times 1.05) = \underline{3780 \text{ kN}}$$

Could use more stiffeners, but likely to be economic to use web thickness = 20 mm

Pier bottom flange

Calculate moments at pier as if main span is built-in at the piers

steel	*say 300 kN × 1/12 × 42*	= *1050 × 1.1*	=	*1155*
concrete	*3.2 × 0.25 × 42 × 26 × 1/12 × 42*	= *3058 × 1.2*	=	*3670*
surfacing	*3.2 × 0.13 × 42 × 24 × 1/12 × 42*	= *1468 × 1.75*	=	*2569*
HB	*say 50% of vehicle as point load*			
	= *1/8 × 1/2 × 1800 × 42*	= *4725 × 1.3*	=	*6143*
				13537 kNm

Assume moment resistance is provided by 90% of elastic resistance of section without web

Lever arm between flanges = 2 m say

Required strength of flange = 13537 / 2 × 1 / 0.9 = 7520 kN

If a flange width of 600 mm is adopted

$$\text{required flange thickness} = \frac{7520 \times 10^3 \times 1.05 \times 1.1}{335 \times 600} = 43 \text{ mm}$$

Adopt pier bottom flange = 600 mm × 50 mm

Commentary to calculation sheet

Specified cover is 10 mm more than the minimum nominal cover prescribed by Clause 4/5.8.2. This is as required by BD 57.

(The chosen slab thickness is the same as that used in SCI publication P066. In that example the cover to reinforcement was assumed to be 5 mm more than the nominal cover specified by BS 5400–4. The increased cover now used reduces the 'lever arm' of the tensile bars in local bending by 5 mm; this is likely to be acceptable.)

Clause 3/6.5.3.1 gives the k-factor as:

$$k = k_d \times k_g \times k_o \times k_s$$

For parts under significant tension (> 0.5 σ_y) with no stress concentrations or sudden loading, the four sub-factors and their product are all equal to 1.0.

Table 3c does not include the effect of reduced yield strength for thicker elements. For material between 40 and 63 mm thick, yield strength is 335 N/mm²; using this value in the expression in 3/6.5.4 would give a limiting thickness of 59 mm.

Preliminary sizing (cont'd)

Pier top flange
Based on experience of similar configurations, the top flange area was chosen to be two-thirds size of bottom flange

Adopt pier top flange = 500 mm × 40 mm

Midspan bottom flange
As for pier, calculate moment as if main span is built-in at the piers. The moment for UDL (steel, concrete, surfacing) at midspan is half that at the pier whereas, for a point load (HB), the moments at midspan and pier are equal

$$\text{Hence, } BM = \tfrac{1}{2}\,(1155 + 3670 + 2569) + 6143 = 9840 \text{ kNm}$$

Lever arm between flanges $= 1.2\,[\text{beam depth} + \tfrac{1}{2} \times 0.25\ (\text{slab})]$
$= 1.33$ m

Required strength of flange $= 9840 / 1.33 = 7400$ kN

Maintaining the same flange width as at pier (600 mm)

$$\text{Required flange thickness} = \frac{7400 \times 10^3}{335 \times 600} \times 1.05 \times 1.1 = 43 \text{ mm}$$

Adopt midspan bottom flange = 600 mm × 45 mm

Midspan top flange
Adopted midspan top flange $= \dfrac{500 \text{ mm} \times 30 \text{ mm}}{(56\% \text{ of bottom})}$

Deck slab
Usual range 230 to 260 mm
Selected slab thickness = 250 mm
Minimum cover to reinforcement: top = 40 mm, bottom = 45 mm

4. STEEL SUB-GRADE

Steel grade S355J2 (to BS EN 10025) has a minimum C_v *of 27 J at −20 °C. The limiting thickness at a minimum effective bridge temperature of −15 °C for parts where k = 1 is thus 55 mm from 3/Table 3c*

All girder material is less than 55 mm thick, so can use steel grade S355J2 generally

Commentary to calculation sheet

Girder make-up and configuration of haunches.
Note the small haunches that accommodate crossfall and facilitate regulation.

The slab reinforcement is T25 at 150 mm centres top and bottom over supports 2 and 3, T20 at 150 mm centres in the side spans and T16 at 150 mm centres in the main span.

Transverse bars are T20 at 150 mm centres; these are placed as the outer layers, with the longitudinal bars between them.

Silwood Park, Ascot, Berks SL5 7QN
Telephone: (01344) 623345
Fax: (01344) 622944

CALCULATION SHEET

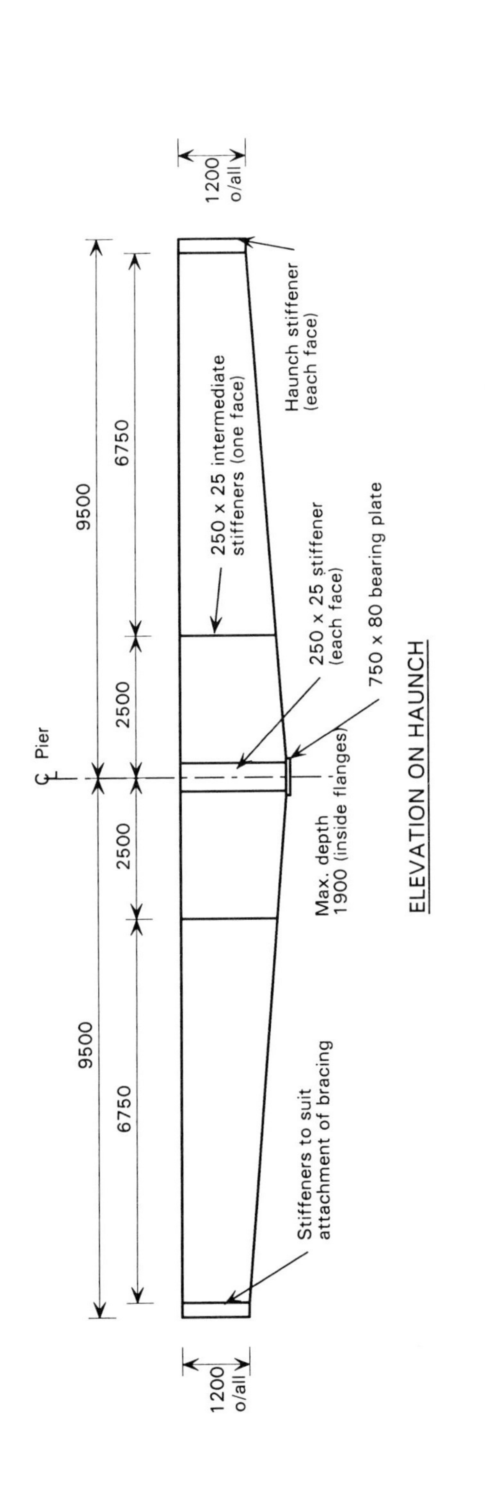

ELEVATION ON HAUNCH

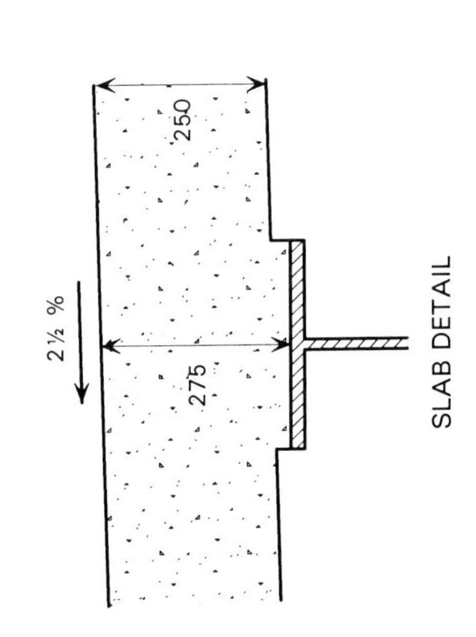

SLAB DETAIL

	Abutment span	*Haunches*	*Main span*
Top flange	*500 × 30*	*500 × 40*	*500 × 30*
Web	*1130 × 16*	*1110 to 1900 × 20*	*1125 × 16*
Bottom flange	*600 × 40*	*600 × 50*	*600 × 45*

GIRDER MAKE-UP

Commentary to calculation sheet

Line-beam models were used for calculation of moments and shears due to dead and superimposed loads. Node positions were chosen to coincide with nodes in the grillage model.

Silwood Park, Ascot, Berks SL5 7QN
Telephone: (01344) 623345
Fax: (01344) 622944

CALCULATION SHEET

Job No:	BCR825	Sheet 6 of 43	Rev B
Job Title	Design of composite bridges – Worked example no. 2		
Subject	Analysis models		
Client	SCI	Made by DCI	Date Sep 2000
		Checked by CRH	Date Dec 2000

5. ANALYSIS MODELS

From previous experience, the tensile stress calculated using uncracked sections in the analysis will exceed 0.1 f_{cu} at the piers

Therefore cracked section properties will be used over 15% of the span either side of the piers,
i.e. over 6.3 m in the main span
over 3.6 m in the side spans

(a) Line beam models

Various line beam models were generated on the computer to analyse the following load cases:

1. 5 no. models with varying lengths of deck concreted (and given long-term section properties) to analyse the concreting loads (see construction sequence cases 3 to 7)

2. Completed deck with long-term composite section properties to analyse the effect of superimposed dead load, differential settlement and the effect of the 'released moments' due to shrinkage modified by creep

3. Completed deck with short-term composite section properties to analyse the effect of the 'released moments' due to the negative and positive differential temperature distributions

(b) Influence line models

A computer program was used to generate influence lines for shear and moment at 11 locations as well as support reactions for a pier and an abutment. The model was as follows (node spacings as for the grillage model, see next page):

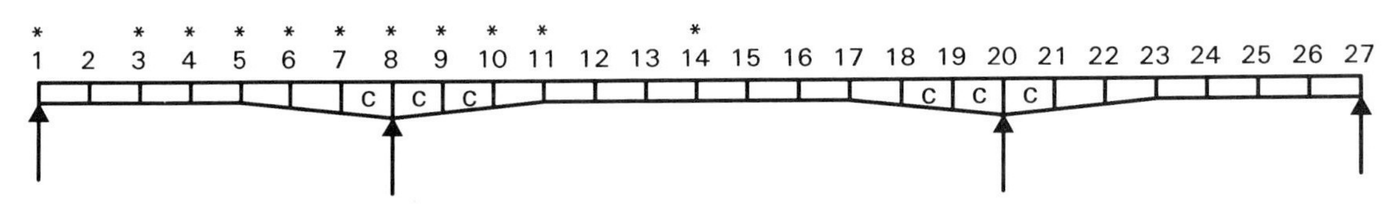

The program used did have a facility for generating influence lines of forces due to vehicles of particular axle spacings and loads. However, it was found just as efficient (and an aid to understanding deck behaviour) to draw the influence lines and sketch the axle spacings for the various HB and abnormal vehicles to scale on sections of tracing paper that could be placed on the influence line graphs

Commentary to calculation sheet

The grillage analysis was carried out before the transverse bracing was designed, and it was chosen to omit the transverse bracing entirely from the model. The bracing was then designed on the basis of the deflections (at the appropriate positions) of the unbraced model, which is conservative. The effect of the bracing on global behaviour is to spread load and again it was considered conservative to omit it.

(c) Grillage model

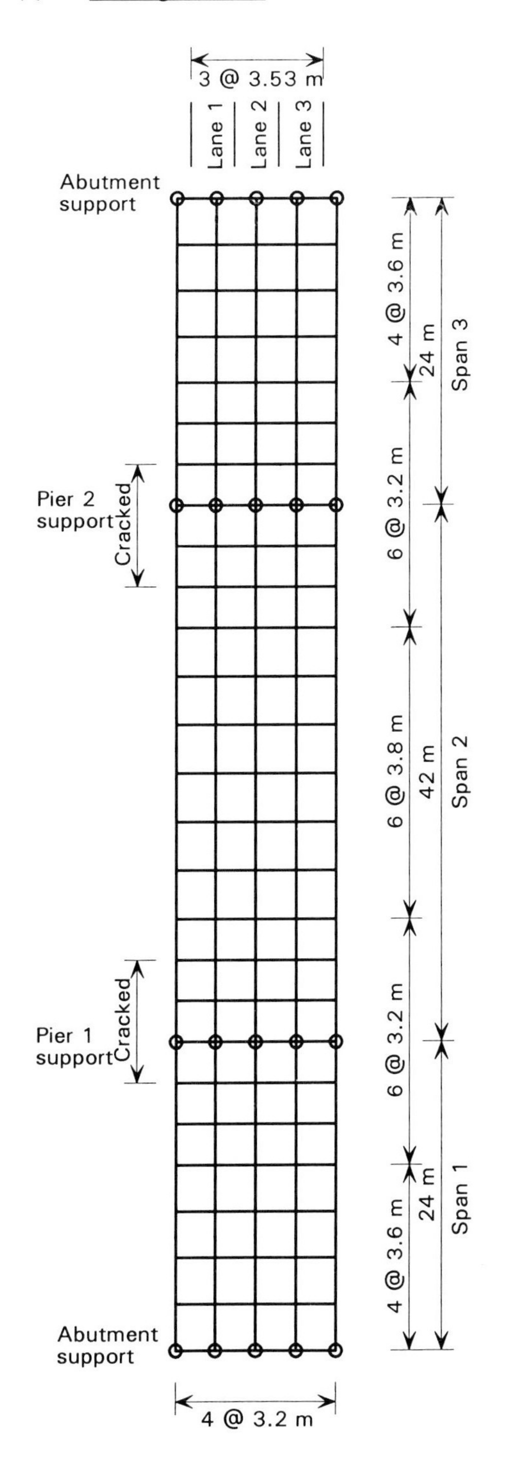

Features of selected grillage model:

1. Nodes at supports and where there is a step change in section properties (i.e. at ends of haunches)

2. Grillage mesh approximately square

3. Uniform node spacing used where possible to facilitate computer generation of the model

4. Nodes required at the centre of the main span

5. 15% of the main span is 6.3 m. The first two members in the main span are given cracked section parameters giving 6.4 m cracked

6. 15% of side span is 3.6 m. The first member in the side span is given cracked section parameters giving 3.2 m cracked

7. Abutment transverse members represent full depth concrete end diaphragms

8. Pier transverse members represent plate diaphragms acting compositely with reinforced concrete slab

9. Remaining transverse members represent 250 mm reinforced concrete slab

10. Cross bracing at haunch ends is assumed to have a negligible effect on the transverse properties of the deck, hence they were omitted

11. Section properties for haunch members calculated at centre of member

Commentary to calculation sheet

Calculation of section properties is given here only for midspan and pier regions, and for gross sections. Additional key values are tabulated on Sheet 9.

Girder make-up is shown on Sheet 5.

It could be argued that there should be a check on the effective thickness of the web, because there are no longitudinal stiffeners to be judged "in accordance with 9.11.5", as required by Clause 9.4.2.5.2. However, because the web of a panel with non-parallel flanges must be treated as a web of a beam with longitudinal stiffeners (Clause 3/9.9.2.1), it may be taken that the check of the combined effects on the web panel is sufficient and the full web thickness may be used, irrespective of y_c/t_w. *In the present case, the depth in the compressive region does not exceed* $68t_w$ *in any case.*

The modulus of elasticity of reinforcement is 200 kN/mm^2 *(see 4/4.3.2), slightly less than that of structural steel (205 kN/mm*^{2}*), so the effective area is reduced.*

Silwood Park, Ascot, Berks SL5 7QN
Telephone: (01344) 623345
Fax: (01344) 622944

CALCULATION SHEET

Job No: *BCR825*		Sheet *8* of *43*	Rev *B*
Job Title *Design of composite bridges – Worked example no. 2*			
Subject *Section properties*			
Client *SCI*	Made by *DCI*	Date *Sep 2000*	
	Checked by *CRH*	Date *Dec 2000*	

6. *SECTION PROPERTIES (Inner beam)*

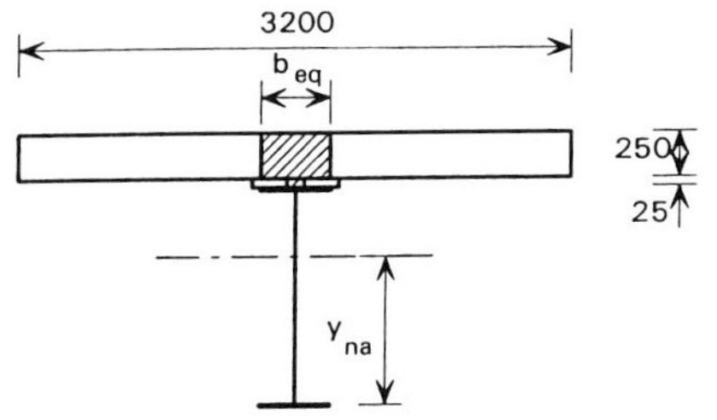

		Short term	Long term
Modular ratios:		205/31 = 6.6	205/15.5 = 13.2
Equivalent slab			
Widths (b_{eq})	3200 slab	485	242
	600 haunch	91	45

Main span midspan

(1) *Steel only*

	A (mm²)	y (mm)	Ay (10³ mm³)	Ay² (10⁶ mm⁴)	I (10⁶ mm⁴)
TF 500×30	15000	1185	17775	21063	1
W 16×1125	18000	607.5	10935	6643	1898
BF 600×45	27000	22.5	607	14	5
	60000		29317	27720	1904

$$I = 27720 + 1904 - \left(\frac{29317^2}{60000}\right) = 15299 \quad \therefore I = \underline{15.30 \times 10^9 \text{ mm}^4}$$

$$y_{na} = 29317 \times 10^3 / 60000 = \underline{489 \text{ mm}}$$

(2) *Composite short*

	A	y	Ay	Ay²	I
Slab 485×250	121250	1350	163688	220978	632
Haunch 91×25	2275	1212.5	2758	3345	0
Beam	60000	489	29318	27720	1904
	183525		195763	252043	2536

$$I = 252043 + 2536 - \left(\frac{195763^2}{183525}\right) = 45761 \quad \therefore I = \underline{45.76 \times 10^9 \text{ mm}^4}$$

$$y_{na} = 195763 \times 10^3 / 183525 = \underline{1067 \text{ mm}}$$

(3) *Composite long* – similar calculations give $y_{na} = \underline{924 \text{ mm}}$, $I = \underline{38.08 \times 10^9 \text{ mm}^4}$

Pier (deep end of haunch)

(1) *Steel only*

	A	y	Ay	Ay²	I
TF 500×40	20000	1970	39400	77618	3
W 20×1900	38000	1000	38000	38000	11432
BF 600×50	30000	25	750	19	6
	88000		78150	115637	11441

$$I = 115637 + 114411 - \left(\frac{78150^2}{88000}\right) = 57675 \quad \therefore I = \underline{57.67 \times 10^9 \text{ mm}^4}$$

$$y_{na} = 78150 \times 10^3 / 88000 = \underline{888 \text{ mm}}$$

(2) *Composite cracked*

Area of one layer of T25s = 21 bars × 12.5² × π × (200 / 205) = 10057 mm²

	A	y	Ay	Ay²	I
T layer (y=1990+275−40−20−½×25)	10057	2192.5	22050	48344	0
B layer (y=1990+25+45+20+ ½×25)	10057	2092.5	21044	44035	0
Beam	88000	888	78150	115637	11441
	108114		121244	208016	11441

$$I = 208016 + 114411 - \left(\frac{121244^2}{108114}\right) = 83488 \quad \therefore I = \underline{83.49 \times 10^9 \text{ mm}^4}$$

$$y_{na} = 121244 \times 10^3 / 108114 = \underline{1121 \text{ mm}}$$

Commentary to calculation sheet

To determine stresses in reinforcement and concrete, use section moduli in appropriate units – multiply the steel modulus by the ratio of the elastic moduli.

The Steel Construction Institute

Silwood Park, Ascot, Berks SL5 7QN
Telephone: (01344) 623345
Fax: (01344) 622944

CALCULATION SHEET

Summary of elastic section properties

N.B. All section properties are given in steel units

Position	State of section	I 2nd moment of area (10^9 mm^4)	y_{na} NA depth (mm)	Section moduli: Top of steel beam (10^6 mm^3)	Section moduli: Bottom of steel beam (10^6 mm^3)	Section moduli: Top of deck* (10^6 mm^3)	Area of section (10^3 mm^2)
Node 1	Steel only	14.69	511	21.3	28.7		57.1
	Short term	42.65	1083	365.3	39.4	108.9	180.6
	Long term	35.76	946	140.7	37.8	67.6	118.8
	Shrinkage	33.53	899	111.5	37.3	58.2	106.5
Node 5L	Steel only	14.69	511	21.3	28.7		57.1
	Cracked	22.15	666	41.5	33.3	30.0	70.0
Node 5R	Steel only	18.51	523	27.4	35.4		72.2
	Cracked	29.38	704	59.2	41.7	42.1	92.3
Node 8	Steel only	57.67	888	52.3	64.9		88.0
	Cracked	83.49	1121	96.1	74.4	77.9	108.1
	Cracked $\psi = 0.73$	78.25	1074	85.4	72.9	70.0	103.3
	Short term	138.57	1618	372.2	85.7	214.1	211.5
	Long term	114.66	1403	195.4	81.7	133.1	149.8
Node 9	Steel only	46.29	802	46.1	57.7		84.3
	Cracked	68.06	1025	87.1	66.4	69.2	104.4
Node 14	Steel only	15.30	489	21.5	31.3		60.0
	Short term	45.76	1067	343.3	42.9	112.1	183.5
	Long term	38.08	924	138.1	41.2	69.2	121.8
	Shrinkage	35.62	876	110.1	40.6	59.5	109.4
Diaphragm	Steel only	20.21	840	24.1	24.1		57.0
	Cracked	21.98	871	27.2	25.2	21.2	58.8
	Short term	40.6	1192	83.2	34.1	53.2	88.8
Node 11	Cracked	20.73	593	34.1	35.0	25.5	68.2

Moduli of elasticity:

$E_{steel} = 205$ kN/mm^2

$E_{reinft} = 200$ kN/mm^2

$E_{conc\ short} = 31$ kN/mm^2

$E_{conc\ long} = 15.5$ kN/mm^2

$E_{conc\ shrink} = 12.4$ kN/mm^2

Girder depth 1806 mm (Node 9)

See Sheet 39 for details of diaphragm

* *To level of top layer of longitudinal reinforcement in cracked sections*

Commentary to calculation sheet

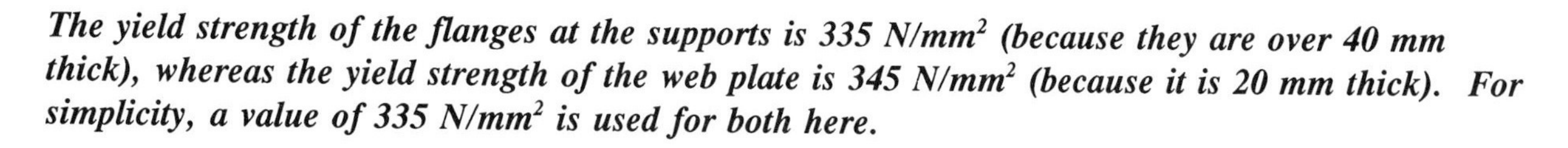

The yield strength of the flanges at the supports is 335 N/mm² (because they are over 40 mm thick), whereas the yield strength of the web plate is 345 N/mm² (because it is 20 mm thick). For simplicity, a value of 335 N/mm² is used for both here.

The area of the reinforcement has been increased in the ratio 0.87 × 460 / (335 / 1.05).

The classification of compactness for webs is given by Clause 3/9.3.7.2. The clause gives limiting overall depth for values of web thickness and proportion of depth that is in compression.

The area of the slab has been reduced in the ratio 0.4 × 40 / (335 / 1.05).

PLASTIC SECTION PROPERTIES AND CLASSIFICATION

Composite section at pier (node 8)

Try web NA at top of web:

	A (mm^2) (comp +)	y (mm)	Ay (mm^3)
Top T25	-12930	242.5	-3,136
Bottom T25	-12930	142.5	-1,843
TF 500 × 40	-20000	20	-400
W 20 × 1900	38000	-950	-36,100
BF 600 × 50	30000	-1925	-57,750
	22140		-99,229
Move NA down by 553.5	-22140	-276.7	6,126
		Z_{pe} =	-93,103
		Y_{pl} =	1397

Proportion of web in compression $m = 1397 / 1900 = 0.709$

Limit for compact section $= 374t_w / (13m - 1)\sqrt{(355/\sigma_y)}$

$= 924$ mm < 1310 mm ∴ Non-compact

Composite section at mid main span (node 14)

Try web NA at top of web

	A (mm^2) (comp +)	y (mm)	Ay (mm^3)
Slab 3200 × 250	40119	180	7,221
Haunch 500 × 25	627	42.5	27
TF 500 × 30	15000	15	225
W 16 × 1125	-18000	-562.5	10,125
BF 600 × 45	-27000	-1147.5	30,983
	10746		48,581
Move NA up by 10.7	-10746	5.4	-58
		Z_{pe}	48,523
		Y_{pl}	1181

PNA is in the top flange ∴ section is compact

Summary of plastic section properties:

		Z_{pe} ($10^6 mm^3$)	Y (mm)	σ_y	M_{pe} (kNm)
Node 1	Steel only	26.6	324	345	9,180
	Composite	45.0	1182	345	15,500
Node 5	Steel only	26.6	324	345	9,180
	Composite cracked	-39.1	826	345	-13,500
Node 8	Steel only	65.5	750	335	21,900
	Composite cracked	-93.1	1397	335	-31,200
Node 14	Steel only	27.3	233	335	9,150
	Composite	48.5	1181	335	16,200

Commentary to calculation sheet

8. ENVIRONMENTAL AND OTHER EFFECTS

(a) *Wind*

Maximum wind gust speed $v_c = v\ k_1\ s_1\ s_2$ — 2/5.3.2.1

where: v = *mean hourly wind speed* = 29 m/s — 2/F2

K_1 = *coefficient related to return period*
= 1.0 *for* 120 *years* — 2/5.3.2.1.2
= 0.85 *for erection*

S_1 = *funnelling factor* = 1.0 *as on open ground* — 2/5.3.2.1.3

S_2 = *gust factor, which varies with loaded length and height above ground level (say 10 m)* — 2/T2
= 1.53 *for a 40 m loaded length*

Wind is a relatively minor load effect on this bridge, but needs to be considered for:

(1) *Fixed and guided bearing horizontal loads (transverse, longitudinal wind)* — 2/5.3.3, 5.3.4

(2) *Steelwork erection, because the beams are transversely weak and prone to lateral torsional buckling, which is made worse by a transverse wind force*

Transverse wind load $P_t = q\ A_1\ C_D$ — 2/5.3.3

where:
A_1 = *solid area facing wind*
C_D = 2.2 *(plate girder) or* 1.28 *(concrete deck)* — 2/5.3.3.2.4
q = $0.613\ v_c^2$ N/m² (v_c in m/s)
v_c = 29 × 0.85 × 1.0 × 1.53 = 37.7 m/s *(erection 40 m loaded length)*
P_t = 1.92 kN/m² *(plate girder erection) or* 1.55 kN/m² *(in service)*

(b) *Shrinkage modified by creep*

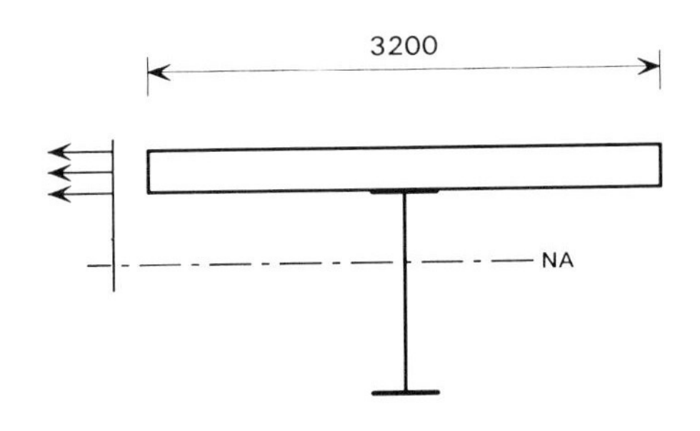

Say conditions are "generally in the open air"
shrinkage strain $\varepsilon_{cs} = -200 \times 10^{-6}$
creep reduction factor $\psi_c = 0.4$
modular ratio $\alpha_e = 205/31 \times 1/0.4 = 16.5$

Effects of shrinkage:

Restrained stress = $200 \times 10^{-6} \times 0.4 \times 31 \times 10^3$ = -2.48 N/mm²

force = $2.48 \times (250 \times 3200 + 500 \times 25)\ 10^{-3}$ = 2015 kN

Restraint moment on the cross section depends on the level of the NA, which varies along the bridge. For the case of the midspan section:

moment = $2.48 \times 10^{-6} \times [250 \times 3200 \times (1350 - 876) + 500 \times 25 \times (1212.5 - 876)]$ = 951 kNm

Commentary to calculation sheet

Primary stresses due to shrinkage are calculated in a similar manner at each different section along the bridge (mean values are used for the haunch sections). The release moments for each section are all applied to a line-beam model to determine the secondary moments, shears and reactions imposed by continuity.

In this case the two bending moment diagrams are:

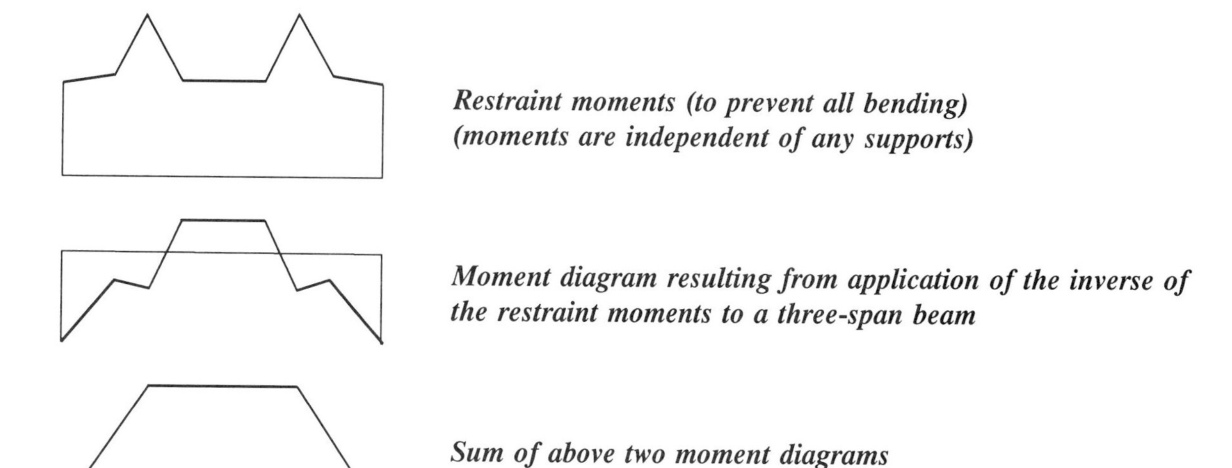

Temperature difference, Clause 2/5.4.5 and 2/C, Table 23.

Primary and secondary effects of differential temperature are calculated in a similar manner to those for shrinkage. Because combination 1 governs the design of the worked example, the calculations are not presented here. Note, however, that the primary stresses in the slab due to positive differential are 2.38 / -0.57 / -0.65 N/mm^2 at the top, 0.6 depth and bottom of the slab respectively (these values are needed for design of shear connection at the abutment).

(Secondary moment = 700 kNm across span 2.)

Minimum and maximum temperature, Clause 2/5.4.3.

In the determination of movement range for the design of the bearings, shrinkage and bending due to live loads will need to be considered in addition to temperature.

(b) *Shrinkage modified by creep (cont'd)*

Values for primary stresses are only likely to be needed for SLS checks, if any
Calculations for the midspan beam give the following primary stress distribution:

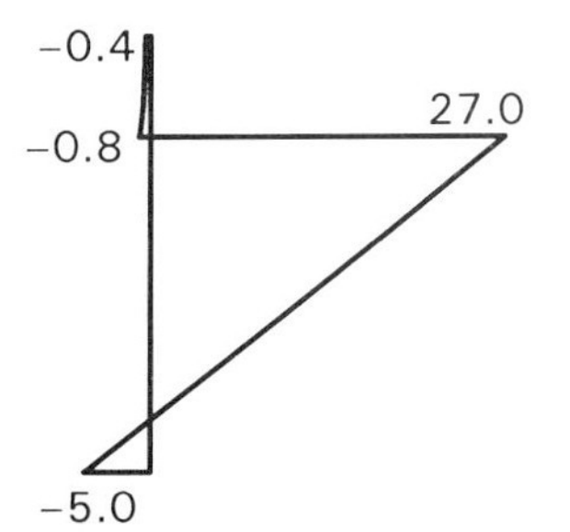

(All stresses in N/mm²)

Secondary moments are calculated in two stages:

(i) Application of 'restraint moments' to a free beam, to keep it straight
(ii) Removal of restraint moments on a model of the three-span composite beam.

(c) *Differential temperature*

(i) *Positive differential*

T_1 = 12.7 °C (250 mm slab, 130 mm surfacing)
$h_1 = 0.6 \times 250 = 150$ mm, $h_2 = 400$ mm

(ii) *Negative differential*

T_1 = -3.5 °C
$h_1 = 150$, $h_2 = 400$ mm

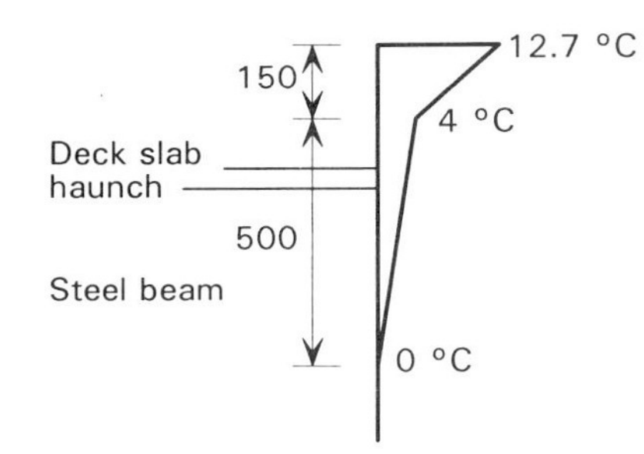

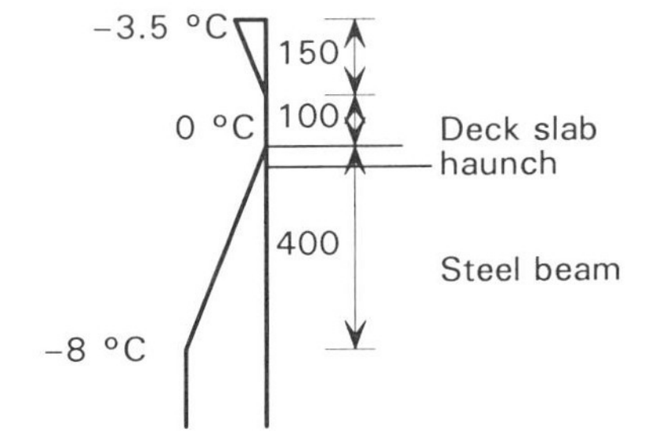

(d) *Temperature*

Minimum effective bridge temperature = –15 °C
Maximum effective bridge temperature = 39 °C – 1 °C = 38 °C
  range of ± 26.5 °C
However, allow bearing to be set at 11.5 ± 10 °C
  range of ± 36.5 °C

2/5.4.6

Fixed at pier → movement = $36.5 \times 0.000012 \times 66000$
= ± 29 mm

Commentary to calculation sheet

The results given are for the effects of unfactored (nominal) loads on 'internal' beams. Appropriate load factors, taken from Part 2, are included in the calculations on subsequent sheets.

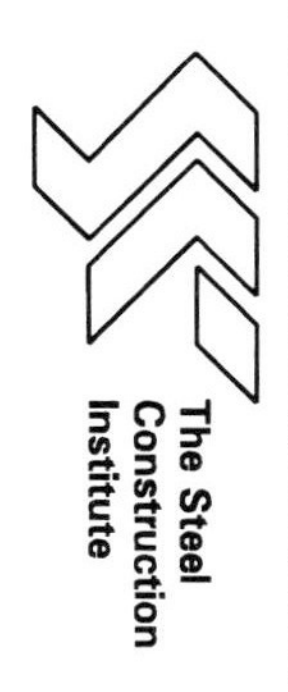

Silwood Park, Ascot, Berks SL5 7QN
Telephone: (01344) 623345
Fax: (01344) 622944

CALCULATION SHEET

9. ANALYSIS RESULT SUMMARY

All forces relate to internal beams, except where noted *R = main span side of pier*

Force	*Reaction (kN)*			*Shears (kN)*				*Moments (kNm)*							
Node	*1*	*8*	*8*	*5*	*8R*	*8R*	*11*	*4*	*5*	*8*	*8*	*9*	*11*	*11*	*14*
Remarks / Loadcase			*Extnl beam*	*Max.*	*Max.*	*Coext with 8 M_{max}*	*Max.*	*Max.*	*Max.*	*Max.*	*Coext with 8R S_{max}*	*Max.*	*Max.*	*Coext with 8 M_{max}*	*Max.*
Steel	*35*	*215*	*215*	*-35*	*115*	*115*	*45*	*-35*	*-125*	*-785*	*-785*	*-440*	*0*	*0*	*240*
Conc. (steel)	*140*	*930*	*930*	*-160*	*500*	*500*	*270*	*1100*	*755*	*-3875*	*-3875*	*-2340*	*-95*	*-95*	*1725*
Conc. (composite)				*-70*				*-1150*	*-1440*	*0*	*0*	*-20*	*50*	*50*	*-200*
Surfacing	*75*	*535*	*208*	*-110*	*290*	*290*	*160*	*80*	*-205*	*-1935*	*-1935*	*-1040*	*305*	*305*	*1220*
Other SDL	*0*	*0*	*264*	*0*	*0*	*0*	*0*	*0*	*0*	*0*	*0*	*0*	*0*	*0*	*0*
HA	*-*	*950*	*576*	*-*	*585*	*-*	*365*	*1855*	*-1480*	*-3180*	*-*	*-*	*-550/ +1345*	*-*	*2615*
HB 45 units	*1007*	*-*	*-*	*-*	*1175*	*-*	*715*	*2660*	*-2290*	*-4790*	*-*	*-*	*-965/ +2540*	*-*	*4440*
AIL	*-*	*1930*	*-*	*-210*	*1235*	*750*	*700*	*3315*	*-2490*	*-5875*	*-2975*	*-3595*	*-875/ +2565*	*1800*	*4505*
Footway	*3*	*15*	*56*	*-5*	*10*	*10*	*10*	*35*	*-50*	*-105*	*-105*	*-35*	*-15/ +25*	*20*	*80*
Settlement	*±4*	*±9*	*±9*	*±4*	*±5*	*5*	*±5*	*±44*	*±58*	*±96*	*-96*	*-82*	*±53*	*±53*	*±29*
Shrinkage/creep	*-55*	*55*	*55*	*-55*	*0*	*0*	*0*	*-550*	*-730*	*-1270*	*-1270*	*-1270*	*-1270*	*-1270*	*-1270*
HB 30 units	*-*	*-*	*-*	*-*	*-*	*-*	*-*	*-*	*-*	*-3870*	*-*	*-*	*-*	*-*	*-*

Commentary to calculation sheet

All the checks on this and subsequent sheets are for the ultimate limit state, unless noted otherwise.

The choice of combination 1 as the governing combination is easily made by inspection.

Checks for erection conditions are given separately on Sheets 40 and 41.

<u>*Check of bending resistance:*</u>
All the beams in this example are without longitudinal stiffeners and should therefore be checked for bending resistance in accordance with Clause 3/9.9.

For the midspan region, where LTB is not a consideration in the completed condition, the first check is made on the moment resistance. Then the check is made on the accumulated stress resulting from construction in stages. The total stress must be less than (factored) yield.

Even though primary shrinkage stress in the top flange is adverse, it does not need to be included because Clause 3/9.9.7 does not require it.

The beam section is compact, which would allow design on the basis of the plastic moment resistance. If that were used, the check on accumulated stress must be at SLS and the effects of differential temperature (primary and secondary effects) and the primary stresses due to shrinkage must be taken into account.

If the elastic moment resistance is used (i.e. designed as a non-compact section), the check on accumulated stress is carried out at ULS. A further check at SLS is not needed here because the shear lag is not large.

The check on concrete stress is made on Sheet 19.

To check the section at the support, the LTB slenderness will need to be calculated. Before embarking on this, the check on accumulated stress is made, to ensure that this is below factored yield stress.

10. MOMENT CHECK (INTERNAL BEAMS)

The forces due to wind, temperature effects, secondary live loads and bearing friction are relatively small with respect to main beam design. Combination 1 governs

Node 14 Main span midspan (kNm)

BM steel only	=	1.05 × 240 (steel) + 1.15 × 1725 (conc)	=	2240
Composite long	=	−1.15 × 200 (conc) + 1.75 × 1220 (SDL) + 1.2 × 29 (settlement)	=	1940
Composite short	=	1.5 × 80 (footway) + 1.3 × 4505 (AIL)	=	5980
		Total moment	=	10160

Shrinkage secondary BM = −1270 kNm, γ_{fL} = 1.2 – *not adverse, so neglect*

Compression flange is restrained by the deck slab, so $\ell_e = 0$ *and* $M_R = M_{ult}$

M_R *is given by either* M_{pe} = 16200 kNm (Sheet 10)

or $Z_{xt}\sigma_{yt}$ = 42.9 × 335 = 14400 kNm (Z_{xt} on Sheet 9)

$M_D = M_R / \gamma_m \gamma_{f3}$ = 16200 / 1.05 × 1.1 = 14000 kNm (compact)

or 14400 / 1.05 × 1.1 = 12500 kNm (non-compact)

Because non-compact resistance is adequate, consider as non-compact section, which means that accumulated stress needs to be checked at ULS

Top flange stress = (2240/21.5) + (1940/138.1) + (5980/343.3) = 136 N/mm²

Bottom flange stress = −(2240/31.3) − (1940/41.2) − (5980/42.9) = 258 N/mm²

Factored yield stress = 335 / 1.05 × 1.1 = 290 N/mm²

∴ *Bending resistance is satisfactory*

Node 8 Pier (kNm)

BM steel only	=	1.05 × −785 (steel) + 1.15 × −3875 (conc)	=	−5280
Composite long	=	1.75 × −1935 (SDL) + 1.2 × −96 (settlement)	=	−3500
Composite short	=	1.5 × −105 (footway) + 1.3 × −5875 (AIL)	=	−7800
Shrinkage secondary BM	=	−1270 × 1.2	=	−1520
		Total moment	=	−18100

Section is non-compact (see Sheet 10), so check accumulated stress first

Top flange stress = −(5280/52.3) − [(3500+7800+1520)/96.1] = −234 N/mm²

Bottom flange stress = (5280/64.9) + [(3500+7800+1520)/74.4] = 254N/mm²

Accumulated stresses less than $\sigma_y / \gamma_m \gamma_{f3}$ (= 335 / 1.155 = 290 N/mm²) – OK

Reinforcement stress = −(3500+7800+1520) / 77.9 × 200 / 205 = −161 N/mm² OK

Commentary to calculation sheet

Determination of lateral torsional buckling slenderness and consequent limiting compressive stress, Clauses 3/9.6, 3/9.7, 3/9.8.

It is assumed that the cross-bracing, in conjunction with the deck slab, provides an effective discrete lateral restraint in accordance with 3/9.6.4.1.1. The minimum required stiffness is that which gives a deflection under unit load of no more than $\ell_R^3 / 40EI_c$ which, for $\ell_R = 9250$ and $I_c = 9 \times 10^8$, is a deflection of 1.07×10^{-4} mm. Strictly, a check should be made.

The Steel Construction Institute

Silwood Park, Ascot, Berks SL5 7QN
Telephone: (01344) 623345
Fax: (01344) 622944

CALCULATION SHEET

Job No: BCR825	Sheet 17 of 43	Rev B
Job Title: Design of composite bridges – Worked example no. 2		
Subject: Moment check		
Client: SCI	Made by: DCI	Date: Sep 2000
	Checked by: CRH	Date: Dec 2000

The haunch section of the compression flange is restrained by a diaphragm at the pier and by a bracing system at the shallow end of the haunch

Effective length l_e *= 9250 mm*

Slenderness parameter $\lambda_{LT} = \lambda_{LT} = \left(\frac{l_e}{r_y}\right) k_4 \ \eta \ v$

where: $K_4 = 1.0$ *(not symmetrical fabricated section)*

Radius of gyration about the y–y axis r_y *is calculated using an equivalent thickness of tension reinforcement*

Equivalent thickness of tension reinforcement
(T25 @ 150 two layers in slab)

$$t = \frac{3200}{150} \times 2 \times 12.5^2 \ \pi \times \frac{200}{205} \text{ (ratio of moduli)} \times \frac{1}{3200} = 6.4 \text{ mm}$$

$$I_y = \frac{1}{12}(3200^3 \times 6.4 + 500^3 \times 40 + 1900 \times 20^3 + 600^3 \times 50) = 1.879 \times 10^{10} \text{ mm}^4$$

$$r_y = (1.879 \times 10^{10} / 108114)^{1/2} \qquad \text{(section area from Sheet 8)}$$

$$= 417 \text{ mm}$$

$$\lambda_f = \left(\frac{\ell_e}{r_y}\right)\left(\frac{t_f}{D}\right)$$

where: t_f = mean flange thickness $= \frac{1}{2}(40 + 6.4 + 50) = 48.2$ mm

$D = 2192 + \frac{1}{2} \times 25 = 2205$ mm (overall depth)

$$\lambda_f = 0.48$$

Second moment of area of compression flange, I_c

$$I_c = 600^3 \times 50 \times \frac{1}{12} = 9 \times 10^8 \text{ mm}^4$$

Tension flange $I_t = (3200^3 \times 6.4 + 500^3 \times 40)\frac{1}{12} = 1.79 \times 10^{10} \text{ mm}^4$

$I_c / (I_c + I_t) = 0.048 = i$

$\psi_i = 2i - 1 = -0.904$

$v = \{[4i(1 - i) + 0.05\,\lambda_f^2 + \psi_i^2]^{1/2} + \psi_i\}^{-1/2} = 3.135$

To determine value of η *requires values of* M_A*,* M_B *and* M_M

From previous sheet $M_A = -18100$

Moment at node 11 (9.6 m from support), from values on Sheet 15

$M = 1.15 \times -45 + 1.75 \times 305 + 1.5 \times 20 + 1.3 \times 1800 + 1.2 \times -53 + 1.2 \times -1270$

(conc) (SDL) (footway) (AIL) (settlement) (shrinkage)

$= 1260$ kNm

Commentary to calculation sheet

M_{ult} for a non-compact section is the least of Z_{xc} σ_{yc}, Z_{xt} σ_{yt} and Z_{xw} σ_{yw} (Clause 3/9.8). For composite sections it should also be limited by $Z_s \times 0.5 f_{cu} \times \gamma_m$, but this is unlikely to govern. Here Z_{xc} σ_{yc} governs, by inspection (see Sheet 9).

(The inclusion of γ_m in the expression for $Z_s \times 0.5 f_{cu} \times \gamma_m$ is to allow for the fact that M_R will be divided by that factor to give the value of M_D.)

The bending resistance is just adequate.

There is some limited scope for reduction of top flange area and increase of bottom flange area.

The load effects at node 5 relate to the case for maximum hogging and the check is on the length between the end of the haunch and the abutment.

No transverse bracing is provided between the end of haunch and the abutment.

It is assumed that both ends of the length (i.e. at abutment and bracing at the haunch) are stiffly restrained, which means that the denominator of the expression for k_e is very close to unity.

Section properties at node 5 were calculated in a similar manner to those on Sheet 17.

The Steel Construction Institute

Silwood Park, Ascot, Berks SL5 7QN
Telephone: (01344) 623345
Fax: (01344) 622944

CALCULATION SHEET

Coexistent shear at node 8

$V = 1.05\times115 + 1.15\times500 + 1.75\times290 + 1.5\times10 + 1.3\times750 + 1.2\times5 = 2200$ kN
(steel) (conc) (SDL) (foot) (AIL) (settlement)

$M_B = 1260 - 2200 \times (9.6 - 9.25) = 490$ kNm

'Average shear' between nodes 8 and 11 (18100 + 1260) / 9.6 = 2020 kN

Approx. value of M_M = ½ × (2200 – 2020) × (9.25 / 2) = 416 kNm

$M_B/M_A = -0.03$, $M_A/M_M = -44$

Hence, $\eta = 0.73$ (Figure 10b)

Thus $\lambda_{LT} = (9250 / 417) \times 1.0 \times 0.73 \times 3.135 = 51$

$M_{ult} = Z_{xc}\,\sigma_{yc} = 74.4 \times 10^6 \times 335 = 24900$ kNm

M_{pe} = 31200 kNm (see Sheet 10)

Hence parameter for Figure 11 $= 51 \times \sqrt{(335/355) \times (24900/31200)} = 44$

From Figure 11a (for welded beams) $M_R / M_{ult} = 0.86$

$M_D = M_R / \gamma_m \gamma_{f3} = (24900 \times 0.86) / 1.05 \times 1.1 = 18500$ <u>ADEQUATE</u>

<u>Node 5 side span, haunch, end internal beam</u> kNm

BM steel only	= 1.05 × –125 (steel) + 1.15 × 755 (conc)	=	740
Composite long	= 1.15 × –1440 (conc) + 1.75 × –205 (SDL) + 1.2 × –58 (settlement)	=	–2080
Composite short	= 1.5 × –50 (footway) + 1.3 × –2490 (AIL)	=	–3310
Shrinkage secondary BM	= –730 × 1.2	=	<u>–880</u>
			–5530

Top flange stress = 740 / 21.3 – (2080 + 3310 + 880) / 41.5 = –116 N/mm²

Bottom flange stress = –740 / 28.7 + (2080 + 3310 + 880) / 33.3 = 163 N/mm²

Check for slenderness and limiting moment of resistance (3/9.6.4.1.1.1, 3/9.7, 3/9.8)

Take $l_e = l_R = (24000 - 9250) = 14750$ mm

r_y = 418 mm

t_f = 37 mm, D = 1420 mm Hence: $\lambda_F = 0.92$ 3/9.7.2

$I_t = 1.15 \times 10^{10}$ mm⁴, $I_c = 7.2 \times 10^8$

Hence $i = 0.059$ $\psi_i = -0.882$ $v = 2.68$

M_A = –5530 kNm

M_B = 0

Commentary to calculation sheet

Strength of reinforcement and concrete, Clause 5/6.2.3.

Calculation of secondary moments due to differential temperature are not shown in the example. Primary effects are not taken into account at ULS (see 3/9.9.7).

Separate assessment of slab for local and global effects is covered by Clauses 5/6.1.2 and 4/4.8.3.

The design of the deck slab for local effects is not included in the worked example.

The Steel Construction Institute

Silwood Park, Ascot, Berks SL5 7QN
Telephone: (01344) 623345
Fax: (01344) 622944

CALCULATION SHEET

Job No: *BCR825*	Sheet *19* of *43*	Rev *B*
Job Title *Design of composite bridges – Worked example no. 2*		
Subject *Moment check*		
Client *SCI*	Made by *DCI*	Date *Sep 2000*
	Checked by *CRH*	Date *Dec 2000*

UDL due to dead load = $1.15 \times 3.2 \times 0.25 \times 26 + 1.75 \times 3.2 \times 0.13 \times 24 + 1.05 \times 0.057 \times 77$
(conc) *(SDL)* *(steel)*

= *46 kN/m*

$M_M \approx WL^2/8 = 46 \times 14.75^2 \times 1/8 = 1250$ kNm $\qquad M_A / M_M \approx -4.4$

Hence η = 0.61 (Figure 10b)

$\lambda_{LT} = (14750 / 418) \times 1.0 \times 0.61 \times 2.68 = 58$

$M_{ult} = Z_{xc}\, \sigma_{yc} = 33.3 \times 10^6 \times 345 = 11490$ kNm

M_{pe} = 13500 kNm (see Sheet 10)

Hence parameter for Figure 11a = $58 \times \sqrt{(345/355) \times (11490/13500)} = 53$

From Figure 11a, $M_R / M_{ult} = 0.76$

$M_R = 0.76 \times 11490 = 8730$ kNm

$M_D = M_R / \gamma_m\, \gamma_{f3} = 8730 / (1.05 \times 1.1) = \underline{7560 \text{ kNm}}$ ***ADEQUATE***

Deck slab (ULS)

At the ultimate limit state, the resistance of a deck slab to global and local loads combined is deemed to be satisfactory if each of these effects is considered separately

Hence, no need to combine local wheel load effects with the global loads

Check summation of stresses

Strength of reinforcement = $0.87 \times 460 / 1.1\ (\gamma_{f3}) = 364$ N/mm²

Strength of concrete = $0.5 \times 40 / 1.1\ (\gamma_{f3}) = 18.2$ N/mm²

At pier (node 8)

Top rebar stress = $[-(3500 + 7800 + 1520) / 77.9] \times 200 / 205 = \underline{-161 \text{ N/mm}^2}$ ***OK***

At midspan (node 14)

Compression at top of slab in combination 1

$$= \frac{1}{13.2}\left(\frac{1940}{69.2}\right) + \frac{1}{6.6}\left(\frac{5980}{112.1}\right) = \underline{10.2 \text{ N/mm}^2 < 18.2 \text{ N/mm}^2} \quad \underline{\text{OK}}$$

Also, consider concrete stresses for combination 3 loading, including the effects of differential temperature

Secondary BM = *700 kNm*

Comb. 3 Comp. short = 1.25×80 (foot) + 1.1×4505 (AIL) + 700×1.0 (diff. temp)

= *5760 kNm*

This is less than the combination 1 short-term moment, so combination 3 does not govern

Commentary to calculation sheet

Determination of effective breadth of slab, Clauses 3/8.2 and 5/5.2.3.

Secondary moment due to shrinkage has been included in the load effect, so the primary shrinkage strain must also be included. The calculation of the value of the strain due to primary shrinkage stresses is not included in the example.

Calculation of crack width, Clause 4/5.8.8.2. Limiting crack widths, Clause 4/4.1.1. Crack widths need only be calculated at the level of cover specified in Part 4, not the increased cover specified in BD 57 (see BA 57 and 4/5.8.8.2).

The stresses used for determining crack width were based on load effects for 30 units of HB. See further comment in Section 7.3.2 of SCI P289 about the use of 25 or 30 units.

Calculation of strain allowing for the effect of shear lag, Clauses 5/5.2.6 and 3/A.6.

Deck slab calculations for local effects are not included in this example.

Crack control longitudinally in deck slab (SLS)

Effective breadth of flange

For pier: main span b/L $= 1.6 / 42 = 0.038$

$\psi = 0.68$ ($\alpha = 0$ as no flange stiffeners)

For pier: side span b/L $= 1.6 / 24 = 0.067, \quad \psi = 0.52$

ψ at pier $= \frac{1}{2}(0.68 + 0.52) = 0.60$

The concrete is assumed to be cracked

$\psi_{cracked} = 1/3\,(1 - 0.60) + 0.60 = \underline{0.73}$

HA factored $= 1.2 \times -3180 = 3816$

HB factored (30 units) $= 1.1 \times -3870 = 4257$ ∴ *HB governs*

Moments

Comp. long $= 1.2 \times -1935$ (SDL) $+ -96 \times 1.0$ (settlement) $= -2420$ kNm

Comp. short $= -4257$ (HB) $+ -105 \times 1.0$ (f'way) $+ -1270 \times 1.0$ (shrink) $= \underline{-5630}$ kNm

-8050 kNm

Loading mainly UDL, so $\psi = 0.73$ *is applicable*

$I = 78.25 \times 10^9$ mm^4 and $y_{na} = 1074$ mm

Calculate strain at top of deck with locked in strain due to shrinkage of 4.3×10^{-5} *(primary shrinkage)*

$$\varepsilon = +4.3 \times 10^{-5} + \frac{8050 \times (2265 - 1074) \times 10^3}{205 \times 78.26 \times 10^9}$$

$$= 6.41 \times 10^{-4}$$

Top mat reinforcement T25 @ 150 with 40 cover to transverse T20
(Ignore the 10 mm extra cover)

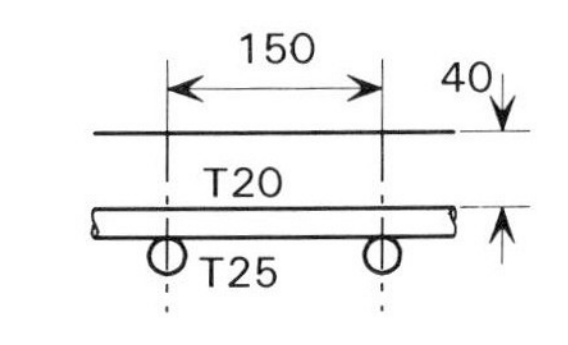

$a_{cr} = (75^2 + 62.5^2)^{1/2} - 12.5 = 85$ mm

Deck slab in overall tension

→ *crack width* $= 3 \times 85 \times 6.41 \times 10^{-4}$

$= \underline{0.16 \text{ mm}} < \underline{0.25}$ <u>*OK*</u>

Crack widths are thus satisfactory over the beams, where there is no longitudinal local bending. Midway between beams, the crack widths due to global bending must be added to those due to local sagging moments. Global strain midway between is given by:

Stress σ_1 *at x from web centreline* $= \sigma_{max}\,[\chi^4 + k\,(1 - \chi^4)]$

where $k = 0.25\,(5\,\psi - 1) = 0.66$

$X = (b - x)/b$ *where b = beam spacing*

Thus, halfway between beams $\sigma_1 = 0.66\,\sigma_{max}$

(Calculations for local effects not included here)

Commentary to calculation sheet

Check of shear resistance and shear/moment interaction:

Beams with parallel flanges and without longitudinal stiffeners are checked in accordance with Clause 3/9.9.

For shear/moment interaction, an effective bending moment is required.

Effective moment is given by the total stress in the extreme fibre (calculated as the sum of stresses for each stage) multiplied by the current section modulus for that fibre (Clause 3/9.9.5.3). The appropriate extreme fibre is the one that governs the value of M_D and M_f.

Although the maximum values of M and V do not occur at the same point in the beam, the interaction check of Clause 3/9.9.3.1 requires the use of the maximum values anywhere in the length of the web panel (between vertical stiffeners) that is being considered.

The Steel Construction Institute

Silwood Park, Ascot, Berks SL5 7QN
Telephone: (01344) 623345
Fax: (01344) 622944

CALCULATION SHEET

11. SHEAR AND SHEAR/MOMENT INTERACTION

Main span centre panel

Over the central 23 m of the main span, the flanges are parallel, so check as a beam without longitudinal stiffeners: — *3/9.9.2.1*

Shear resistance under pure shear $V_D = \dfrac{t_w\,(d_w - h_h)\,\tau_l}{\gamma_m\,\gamma_{f3}}$ — *3/9.9.2.2*

Web thickness $t_w = 16$ *mm*

$d_w - h_h = 1125 - 0 = 1125$ *mm*

To calculate τ_l *requires the following parameters:*

$\tau_y = \sigma_{yw}/\sqrt{3} = 355/1.73 = 205\ \text{N/mm}^2$ — *3/9.9.2.2*

ϕ = *aspect ratio of panel* = 23500 / 1125 = 20.9

$\lambda = d_{we}/t_w = 1125/16 = 70.3$

$$m_{fw} = \frac{\sigma_{yf}\,.\,b_{fe}\,.\,t_f^2}{2\,\sigma_{yw}\,.\,d_{we}^2\,.\,t_{we}} = \frac{345 \times 250 \times 30^2}{2 \times 355 \times 1125^2 \times 16} = 0.005 \text{ (smaller at top flange)}$$

With $m_{fw} = 0.005$, $\tau_l/\tau_y = 0.86$ *With* $m_{fw} = 0$, $\tau_l/\tau_y = 0.85$ — *3/Figure 13*

$V_D = 16 \times 1125 \times 0.86 \times 205 / (1.05 \times 1.1)$ N V_D = <u>*2750 kN*</u>

and $V_R = 16 \times 1125 \times 0.85 \times 205 / (1.05 \times 1.1)$ N V_R = <u>*2720 kN*</u> — *3/Figure 12*

Maximum shear (Node 11) $V = 1.05{\times}45 + 1.15{\times}270 + 1.75{\times}160 + 1.5{\times}10 + 1.3{\times}715 + 1.2{\times}5$

(steel) (conc) (SDL) (foot) (HB) (settle)

V = <u>*1590 kN*</u>

M_D = <u>*12500 kNm*</u> *(Sheet 16)*

For value of M_f:

Height of centroid of composite flange above soffit of beam

$$= \frac{121250 \times 1350 + 2275 \times 1212.5 + 15000 \times 1185}{121250 + 2275 + 15000} = 1330 \text{ mm}$$

$d_f = 1330 - 45 \times ½ = 1307$ *mm*

$$M_f = \frac{f_f\,.\,d_f}{\gamma_m\,.\,\gamma_{f3}} = \frac{(335 \times 600 \times 45) \times 1307}{1.05 \times 1.1} \text{ Nmm}$$

= <u>*10240 kNm*</u>

Effective maximum BM (Node 14)

$M = 42.9 \times 258$ *(Total bottom flange stress from Sheet 16)*

= <u>*11070 kNm*</u>

$V < V_D$ $V < V_R$ $M < M_D$ *but* $M > M_f$

$$\frac{M}{M_D} + \left(1 - \frac{M_f}{M_D}\right)\left(\frac{2V}{V_R} - 1\right) = \frac{11070}{12500} + \left(1 - \frac{10240}{12500}\right)\left(\frac{2 \times 1590}{2720} - 1\right) = 0.92 \text{ Satisfactory}$$

Commentary to calculation sheet

The value of M_R / Z_{xc} *gives a stress of 262 N/mm*^{2}*, so use that value, rather than* σ_y.
(See 3/9.9.3.1.)

The Steel Construction Institute

Silwood Park, Ascot, Berks SL5 7QN
Telephone: (01344) 623345
Fax: (01344) 622944

CALCULATION SHEET

Job No: *BCR825*	Sheet *22* of *43*	Rev *B*
Job Title *Design of composite bridges – Worked example no. 2*		
Subject *Shear and shear/moment interaction*		
Client *SCI*	Made by *DCI*	Date *Sep 2000*
	Checked by *CRH*	Date *Dec 2000*

Side span abutment panel

Flanges are parallel, check as beam without longitudinal stiffeners

By inspection, depth of web in compression y_c is less than 68 × 16 = 1088

Effective web thickness $t_{we} = t_w = 16.0$ mm

Aspect ratio ϕ = 14750 / 1130 = 13.05

$$m_{fw} = \frac{\sigma_{yf} \cdot b_{fe} \cdot t_f^2}{2\,\sigma_{yw} \cdot d_{we}^2\, t_w} = \frac{345 \times 250 \times 30^2}{2 \times 355 \times 1130^2 \times 16} = 0.005$$

$$\lambda = d_{we} / t_w = 1130/16 = 71$$

With $m_{fw} = 0.005$, $\tau_l/\tau_y = 0.88$. *With* $m_{fw} = 0.0$, $\tau_l/\tau_y = 0.86$

$V_D = 16 \times 1130 \times 0.88 \times 205 / (1.05 \times 1.1)$ N V_D = __2820 kN__

and $V_R = 16 \times 1130 \times 0.86 \times 205 / (1.05 \times 1.1)$ N V_R = __2760 kN__

Shear, Node 5 = 1.05×35 (steel) + 1.15×230 (conc) + 1.75×110 (SDL) + 1.5×5 (foot) + 1.3×210 (AIL) + 1.2×4 (settle) + 1.2×55 (shrink)
= 845 kN

Shear, Node 1 = 1.05×35 + 1.15×140 + 1.75×75 + 1.5×3 + 1.3×1007 (HB) + 1.2×4
= 1643 kN

Sagging moment at Node 4

steel only = 1.0 × −35 (steel) + 1.15 × 1100 = 1230 kNm

composite = 1.15× −1150 (conc) + 1.75×80 (SDL) + 1.5×35 (foot) + 1.3×3315 (AIL) + 1.2×44 (settle)
= 3230 kNm

Comparison with Node 5 hogging moments shows that Node 5 is worst case and that bottom (compression) flange is critical

M = 33.3 × 163 = __5430 kNm__ (Stress from Sheet 18)

Height of centroid of composite flange above soffit of beam using cracked section properties (2 layers of 21 no. T20 bars @ 150 centres)

$$\text{Area/layer} = 21 \times 314 \left(\frac{200}{205}\right) = 6433\ \text{mm}^2$$

$$\text{lever arm } d_f = \frac{6433\,(1405 + 1300) + 500 \times 30 \times 1185}{2 \times 6436 + 500 \times 30} - \frac{1}{2} \times 40 = 1242\ \text{mm}$$

Flange force $f_f = M_R / Z_{xc} \times A_f$ (M_R = 8730, Sheet 19)

= 8730 / 33.3 × 600 × 40 = 6290 kN

$$M_f = \frac{f_f \cdot d_f}{\gamma_m \cdot \gamma_{f3}} = \frac{6290 \times 1242}{1.05 \times 1.1}\ \text{Nmm} = \underline{6760\ \text{kNm}}$$

M_D = __7560 kNm__ (Sheet 19)

$M < M_f$ and $V < V_R$ ∴ __Panel OK__

Commentary to calculation sheet

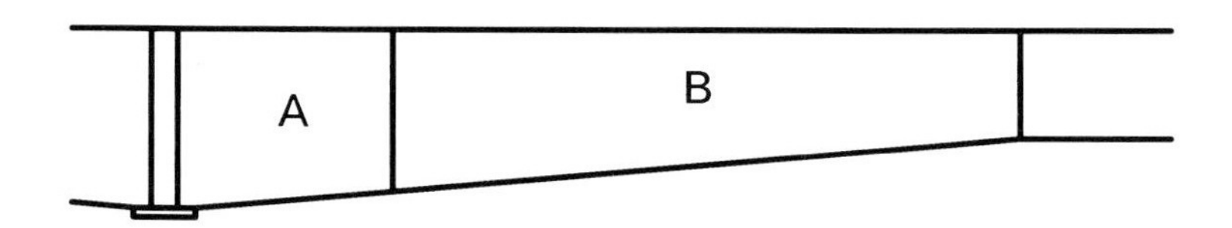

Depth varies by (1990–1200) over a length of (9500–375).

Where the flanges are not parallel, Clause 3/9.9.2.1 requires the shear strength of the panel (and shear/bending interaction) to be checked in accordance with Clause 3/9.11.

Part of the shear is carried out by the inclined force in the compression flange. The shear stress in the web panel is calculated from the net shear force (after deducting this component).

In the calculation shown opposite of net shear in the web, the force in the compression flange is taken as that due to maximum moment; this is non-conservative because the (lower) stress coexistent with maximum shear should have been used, but the difference is small (2 N/mm^2 shear stress). On the other hand, it may also be noted that the longitudinal stresses used in the buckling and yield checks on Sheets 23 and 24 are also those for the maximum moment case; this is conservative. Strictly, all checks should be made with coexistent effects.

Yield of web panels, Clause 3/9.11.3.

Note also that the transverse stress from the top flange has not been included in the yield check, because it is slightly beneficial and in any case has dispersed by this level. It should be included in a yield check on the upper web region, but here that check is deemed satisfactory by inspection.

Main span haunch panel A

Flanges not parallel → design web as for beam with longitudinal stiffeners

Consider panel adjacent to bearing stiffener

$a = 2500 - 160 = 2340$ *mm*

Depth of panel varies – conservatively consider maximum web depth, i.e. $b = 1900$ *mm*

By inspection, maximum bending stresses occur at pier

Consider bending stresses at top and bottom of web

Max. stress in top flange	=	-234 N/mm² (Sheet 16)	
Coexistent in bottom flange	=	254 N/mm²	
Stress at top of web	=	$-234 + (40 / 1990)(234 + 254)$	$= -224$ N/mm²
Stress at bottom of web	=	$254 - (50 / 1990)(234 + 254)$	$= 242$ N/mm²

Max. shear $= 1.05\times115 + 1.15\times500 + 1.75\times290 + 1.5\times10 + 1.3\times1235 + 1.2\times5$

(Node 8R) (steel) (conc) (SDL) (footway) (AIL) (settlement)

$= \underline{2830 \text{ kN}}$

$$\text{Shear in web} = \left[2830\times10^3 - (248\times600\times50)\times\frac{790}{9125}\right]\times\left[\frac{1}{1900\times20}\right]$$

$$= \underline{57 \text{ N/mm}^2}$$

For transverse stress at top of web, consider patch load under wheels

HA wheel load $= 1.5\,(\gamma_{fc}) \times 100 = 150$ *kN*

HB wheel load $= 1.3\,(\gamma_{fc}) \times 112.5 = 146$ *kN*

$\sigma_2 = 150 \times 10^3 / (1068 \times 20) = 7$ *N/mm²*

This stress is conservatively taken all along top edge of panel

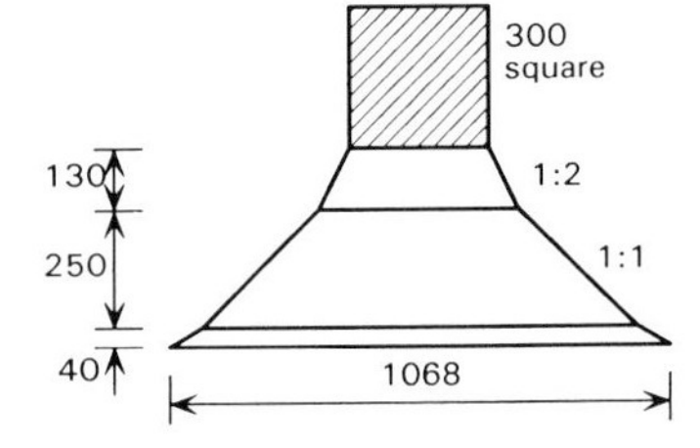

Summary of stresses

$\sigma_1 = +9 \qquad \sigma_{b1} = +233 \qquad \sigma_2 = +7 \qquad \tau = 57$ N/mm²

In compression region (lower web)

$\sigma_{1e} = +9 + 0.77 \times (+233) = +188$ N/mm²

Yield check:

Strength of web panel for yield check $= 345 / (1.1 \times 1.05) = \underline{299 \text{ N/mm}^2}$

In lower web:

$$\sigma_{1e}^2 + \sigma_2^2 - \sigma_{1e}\sigma_2 + 3\tau^2 = 188^2 + 0^2 - (188 \times 0) + 3 \times 57^2 = (212)^2$$

$\underline{212 \text{ N/mm}^2 < 299 \text{ N/mm}^2}$ *OK*

Commentary to calculation sheet

Buckling of web panels, Clause 3/9.11.4.

The use of σ_2 on the tension side of the panel is conservative.

The check opposite combines maximum shear with maximum moment. If the check failed by a small margin, a more detailed check with coexistent values would be made.

Main span haunch panel A (cont'd)

Buckling check:

Aspect ratio, $\phi = a/b = 2340 / 1900 = 1.23$

For k_1, k_q, k_b *restraint depends on* ϕ, M_{fw}, $\lambda = b/t_w = 95$

To be restrained, m_{fw} *should be* > 0.01 *for both flanges (from Figure 22)*

$$(m_{fw})\ top = \frac{335 \times 250 \times 40^2}{2 \times 345 \times 1900^2 \times 20} = 0.0027 < 0.01$$

$\therefore$ *panel is unrestrained for* k_1, k_q *and* k_b

Panel continues beyond transverse stiffeners bounding panel by at least a/2 – hence panel is restrained for k_2

Buckling coefficients:

k_1 *greater of* (a) $\lambda = b/t_w = 95 \rightarrow \underline{k_1 = 0.33}$ *(Curve 2, Figure 23a)*

(b) $\lambda = a/t_w = 117 \rightarrow \underline{k_1 = 0.03}$ *(Curve 3, Figure 23a)*

$k_1 = \underline{0.33}$

k_q $\lambda = 95$ $\phi = 1.0$ $k_q = 0.71$

$\phi = 2.0$ $k_q = 0.64$ *interpolate* $k_q = \underline{0.69}$ *(Figure 23b)*

$k_b =$ $\lambda = 95$ $k_b = 1.04$ *(Figure 23c)*

k_2 *greater of* (a) $\lambda = 117 \rightarrow$ $k_2 = 0.34$ *(Curve 1)*

(b) $\lambda = 95 \rightarrow$ $k_2 = 0.05$ *(Curve 3)* $k_2 = \underline{0.34}$

Interaction coefficients:

$$m_1 = \left(\frac{\sigma_1 \cdot \gamma_m \cdot \gamma_{f3}}{\sigma_{yw} \cdot k_1 (1 - \rho)}\right)^2 = \left(\frac{9 \times 1.05 \times 1.1}{345 \times 0.33 \times 1.0}\right)^2 = +0.008$$

$\rho = 0$ *as no redistribution of stress*

$$m_2 = \left(\frac{\sigma_2 \cdot \gamma_m \cdot \gamma_{f3}}{\sigma_{yw} \cdot k_2}\right)^2 = \left(\frac{7 \times 1.05 \times 1.1}{345 \times 0.34}\right)^2 = +0.005$$

$$m_c = (m_1 + m_2)^{1/2} = 0.114$$

$$m_b = \left(\frac{\sigma_b \cdot \gamma_m \cdot \gamma_{f3}}{\sigma_{yw} \cdot k_b (1 - \rho)}\right)^2 = \left(\frac{233 \times 1.05 \times 1.1}{345 \times 1.04 \times 1.0}\right)^2 = 0.563$$

$$m_q = \left(\frac{\tau \cdot \gamma_m \cdot \gamma_{f3}}{\sigma_{yw} \cdot k_q}\right)^2 = \left(\frac{57 \times 1.05 \times 1.1}{345 \times 0.69}\right)^2 = 0.076$$

$$\therefore m_c + m_b + 3\,m_q = \underline{0.905 < 1.0}$$

Commentary to calculation sheet

Detailed results for shear forces and shear stress calculations are not included in the example.

The Steel Construction Institute

Silwood Park, Ascot, Berks SL5 7QN
Telephone: (01344) 623345
Fax: (01344) 622944

CALCULATION SHEET

Main span haunch panel B

Shear stress in web remains at about 60 N/mm² throughout much of the haunch (the web depth decreases and the component carried by flange also reduces)

If yield criterion is satisfied by panel A, it will also be satisfied by panel B

Buckling check:
Depth of beam at panels A/B intersection

$$= 1200 + (1990 - 1200)\left(\frac{9125 - 2125}{9125}\right) = 1806 \text{ mm}$$

$b = 1806 - 90 = 1716$ mm
$a = 6750$ mm
$\phi = a/b = 3.9$

By inspection, panel unrestrained for k_1, k_b, k_q but restrained for k_2

Buckling coefficients:

k_1 greater of (a) $\lambda = b/t_w = 86 \rightarrow k_1 = 0.36$ (Curve 2, Figure 23a)
(b) $\lambda = a/t_w = 338 \rightarrow k_1 = 0 \quad \rightarrow k_1 = \underline{0.36}$
$k_q \quad \lambda = 86 \quad \phi \geq 3.0 \rightarrow k_q = \underline{0.64}$ (Figure 23b)
$k_b \quad \lambda = 86 \rightarrow k_b = \underline{1.06}$ (Figure 23c)
k_2 greater of (a) $\lambda = 338 \rightarrow k_2 = 0.24$ (Curve 1) (Figure 23a)
(b) $\lambda = 86 \rightarrow k_2 = 0.06 \quad k_2 = \underline{0.24}$

Calculate max. moments at Node 9:

Steel only $= -1.05 \times 440$ (steel) $- 1.15 \times 2340$ (conc) $= \underline{-3150 \text{ kNm}}$
Composite $= -1.15 \times 20$ (conc) $- 1.75 \times 1040$ (SDL) $- 1.5 \times 35$ (footway) $- 1.3 \times 3595$ (AIL) $- 1.2 \times 82$ (settle) $- 1.2 \times 1270$ (shrink)
$= \underline{-8190 \text{ kNm}}$

Max. moments at Node 8 (from Sheet 16):

Steel only $= -5280$ kNm
Composite $= -12820$ kNm

Interpolate to give moment at panel A/B intersection noting that Node 9 is 3200 mm from the pier support whereas the panel A/B intersection is 2500 mm from the pier:

Steel only $= -3150 - (5280 - 3150) \times (3200 - 2500) / 3200 = -3620$ kNm
Composite $= -8190 - (12820 - 8190) \times (3200 - 2500) / 3200 = -9200$ kNm

Stress at top of beam $= -(3620 / 46.1) - (9200 / 87.1) = -184$ N/mm²
Stress at bottom of beam $= +(3620 / 57.7) + (9200 / 66.4) = +202$ N/mm²

Commentary to calculation sheet

Stress at top of web

$$= -184 + (40 / 1806)(184 + 202) = -175 \text{ N/mm}^2$$

Stress at bottom of web

$$= +202 - (50 / 1806)(184 + 202) = +191 \text{ N/mm}^2$$

$\sigma_1 = +8$ $\quad\sigma_b = 183$ $\quad\sigma_2 = 7$ $\quad\tau = 60 \text{ N/mm}^2$

Interaction coefficients:

$$m_1 = \left(\frac{8 \times 1.05 \times 1.1}{345 \times 0.36 \times 1}\right)^2 = 0.006 \qquad m_2 = \left(\frac{7 \times 1.05 \times 1.1}{345 \times 0.24 \times 1}\right)^2 = 0.010$$

$$m_c = (m_1 + m_2)^{1/2} = 0.126$$

$$m_b = \left(\frac{183 \times 1.05 \times 1.1}{345 \times 1.06 \times 1}\right)^2 = 0.334$$

$$m_q = \left(\frac{60 \times 1.05 \times 1.1}{345 \times 0.64}\right)^2 = 0.099$$

$m_c + m_b + 3\, m_c$ = <u>$0.757 < 1.0$</u> <u>*OK*</u>

The above calculation combines maximum moment with maximum shear. Thus, panel B has been shown to be adequate for both the yielding and buckling criteria

<u>*Side span haunch*</u>

Similar calculations to the main span haunch above show that panels A and B in the side span haunch are also adequate in yielding and buckling

Commentary to calculation sheet

Design of transverse web stiffeners, Clause 3/9.13.

In the following checks on the stiffener, no forces have been included due to transverse action of the bracing frame. These should normally be calculated and included, though the stiffener is seen to be quite adequate to take such forces here.

Tension field action is not invoked in the calculation of the shear strength of the web for beams with longitudinal stiffeners. However, Clause 3/9.13.3.2 implies that tension fields do develop in individual web panels (and in unstiffened webs of tapered beams). Vertical stiffeners must therefore be designed to resist any such forces. There are no such forces in this example.

The designer has used the value of 60 N/mm² for shear stress in panel B.
τ_R is the lesser of τ and τ_0 and thus is taken as 60 N/mm² here.

It is conservative to use the smaller value of a in deriving F_{wi}.

The local wheel load is dispersed over a longer length of web than that acting effectively with the stiffener, but it is conservative to consider it acting entirely on the stiffener.

The Steel Construction Institute

Silwood Park, Ascot, Berks SL5 7QN
Telephone: (01344) 623345
Fax: (01344) 622944

CALCULATION SHEET

Job No: BCR825	Sheet 27 of 43	Rev B
Job Title: *Design of composite bridges – Worked example no. 2*		
Subject: *Stiffener design*		
Client: SCI	Made by DCI	Date Sep 2000
	Checked by CRH	Date Dec 2000

12. STIFFENER DESIGN

Main span haunch stiffener between panels A and B

At stiffener, $t_w = 20$ mm, $b = 1716$ mm, $\sigma_1 = 8$ N/mm² (Sheet 26)

$\sigma_b = 183$ N/mm², $\tau = 60$ N/mm²

Loading on stiffener:

(a) *Force due to tension field action. Check whether tension field action occurs*

$$2.9\,E\left(\frac{t_w}{b}\right)^2 = 2.9 \times 205000\left(\frac{20}{1716}\right)^2 = 81\ \text{N/mm}^2 > \sigma_1$$

$$\rightarrow \quad \tau_o = 3.6\,E\left[1 + \left(\frac{b}{a}\right)^2\right]\left(\frac{t_w}{b}\right)^2\left[1 - \frac{\sigma_1}{2.9E}\left(\frac{b}{t_w}\right)^2\right]^{1/2}$$ 3/9.13.3.2

Panel A: $a = 2340$

Take b as the average depth = ½ (1900 + 1716) = 1808 mm

Above equation gives $\tau_o = 136$ N/mm²

Panel B: $a = 6750$

b = ½ (1716 + 1110) = 1413 mm

Above equation gives $\tau_o = 149$ N/mm²

Both values $> \tau$ ∴ *no tension field occurs,* $F_{tw} = 0$

(b) *Web destabilisation force* 3/9.13.3.3

$F_{wi} = l_s^2\, t_w\, k_s\, \sigma_R / a$

a = 2340 mm (conservative)

$\sigma_R = \tau_R + (1 + \Sigma A_s / l_s t_w)(\sigma_1 + 1/6\ \sigma_b)$

but $\Sigma A_s = 0$ as no longitudinal stiffeners

$\sigma_R = 60 + (8 + 1/6 \times 183)$

$= 99$ N/mm²

$l_s = 1716$ mm

$\lambda = l_s / r_{se}\,(\sigma_y / 355)^{1/2} = 22$

$\rightarrow k_s = 0.06$

$\rightarrow F_{wi} = \underline{149\ \text{kN}}$

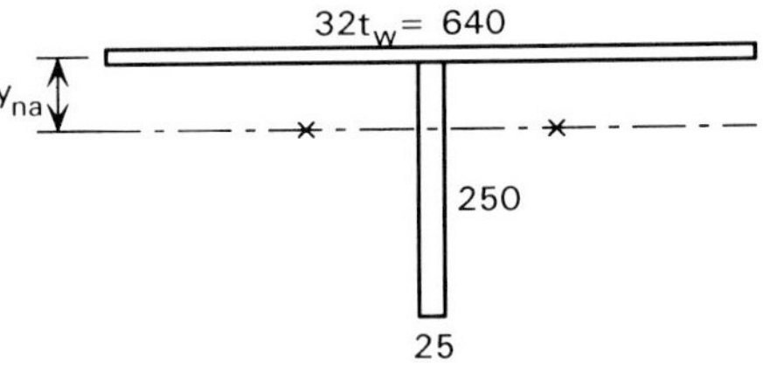

Effective Stiffener Section

$A_{se} = 19050$ mm²

$y_{na} = 44.3$ mm

$I_{se}\,(x\text{-}x) = 1.095 \times 10^8$ mm⁴

$r_{se} = 76$ mm

(c) *Force due to local wheel load*

HA wheel load = 150 kN (Sheet 23) Taken as applied along centreline of web

Check yield of web plate:

Axial force $F_{tw} + F_t = 0 + 150 = 150$ kN (reduces to zero at bottom of web)

Moment $= 150 \times 10^3 \times 44.3 = 6.65$ kNm

$$\sigma_{es2} = \frac{150 \times 10^3}{19050} + \frac{6650 \times 10^3 \times 44.3}{1.095 \times 10^8} \quad \text{(stress at mid-plane)}$$

$= 11$ N/mm²

$\sigma_e = [(\sigma_1 + k\,\sigma_b)^2 + \sigma_{es2}^2 - \sigma_{es2}\,(\sigma_1 + k\,\sigma_b) + 3\,\tau_R^2]^{1/2}$

Top: $\sigma_e = [(8-0.77\times183)^2 + 11^2 - 11\,(8-0.77\times183) + 3\times60^2]^{1/2} = \underline{173 < 299\ \text{N/mm}^2}$

Bottom: $\sigma_e = [(8+0.77\times183)^2 + 0^2 + 3\times60^2]^{1/2} = \underline{181 < 299\ \text{N/mm}^2}$ OK

Commentary to calculation sheet

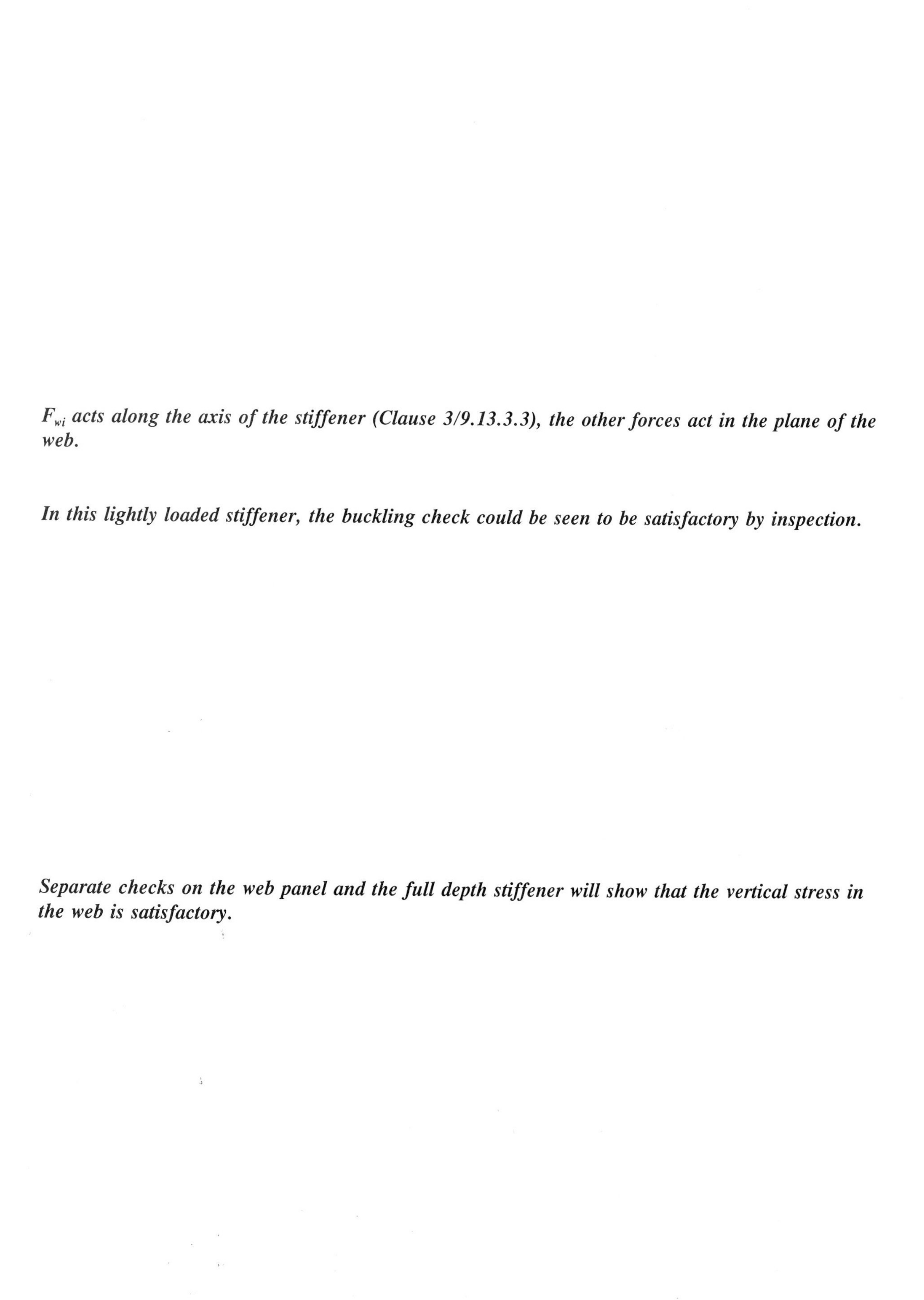

F_{wi} acts along the axis of the stiffener (Clause 3/9.13.3.3), the other forces act in the plane of the web.

In this lightly loaded stiffener, the buckling check could be seen to be satisfactory by inspection.

Separate checks on the web panel and the full depth stiffener will show that the vertical stress in the web is satisfactory.

The Steel Construction Institute

Silwood Park, Ascot, Berks SL5 7QN
Telephone: (01344) 623345
Fax: (01344) 622944

Check yield of stiffener: 3/9.13.5.2

Stress in extreme fibre

$$= \frac{150 \times 10^3}{19050} - \frac{6650 \times 10^3 \times (250 - 34.3)}{1.095 \times 10^8}$$

$= \underline{-5\ N/mm^2}$ <u>OK</u>

Buckling of stiffener: 3/9.13.5.3

$\lambda = l_s / r_{se}\ (\sigma_y / 355)^{½} = 22$

From 3/Figure 37, $\sigma_c / \sigma_y = 0.96$ Hence $\sigma_{ls} = 331\ N/mm^2$

$P = F_{tw} + F_{wi} + F_t = 0 + 149 + 150 = 299\ kN$

$M_{xs} = (F_{tw} + F_t)\ y_{na} = (0 + 150) \times 0.044 = 6.65\ kNm$

Z_x = modulus for compression fibre $= 1.09 \times 10^8 / 54.3$
$= 2.0 \times 10^6$

$$\gamma_m\, \gamma_{f3} \left[\left(\frac{P}{A_{se}\, \sigma_{ls}} \right) + \frac{M_{xs}}{z_x\, \sigma_{ys}} \right] = 1.2 \times 1.1 \left[\frac{299\,000}{330 \times 19050} + \frac{6.65}{2.0 \times 345} \right]$$

$= \underline{0.08 < 1.0}$ <u>OK</u>

<u>Main span haunch stiffener between panels A and B OK in yield and buckling</u>

<u>Stiffener at shallow end of haunch</u> (kink in flange)

Force due to change of slope of bottom flange

$= 163\ N/mm^2 \times 600 \times 40 \times 10^{-3} \times 790 / 9150$

$= 338\ kN$ (Stress at Node 5 from Sheet 18)

Stiffener 250×25 *each side*

Length of web acting $= 2 \times 20 \tan 60° = 69\ mm$

Area acting $= 2 \times 250 \times 25 + 16 \times 69 = 13604\ mm^2$

$\sigma = 338000 / 13604 = \underline{25\ N/mm^2}$ <u>OK</u>

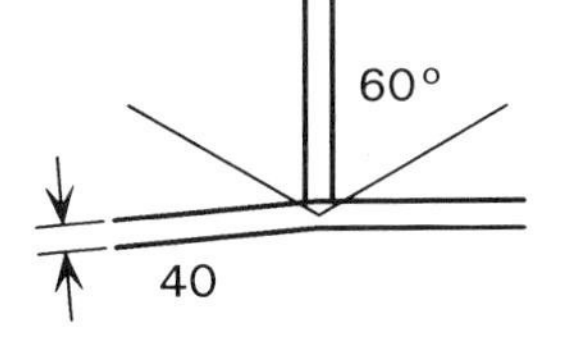

By top of stiffener, length of web acting

$= 69 + 2 \times 300 \tan 45° \text{ but } \ngtr 32\ t_w = 640$

$= 669\ mm$

On one side the web is 20 mm thick, on the other 16 mm thick

Stress $= 338000 / [335 \times (16 + 20)] = 28\ N/mm^2$

Commentary to calculation sheet

Load bearing support stiffeners, Clause 3/9.14.

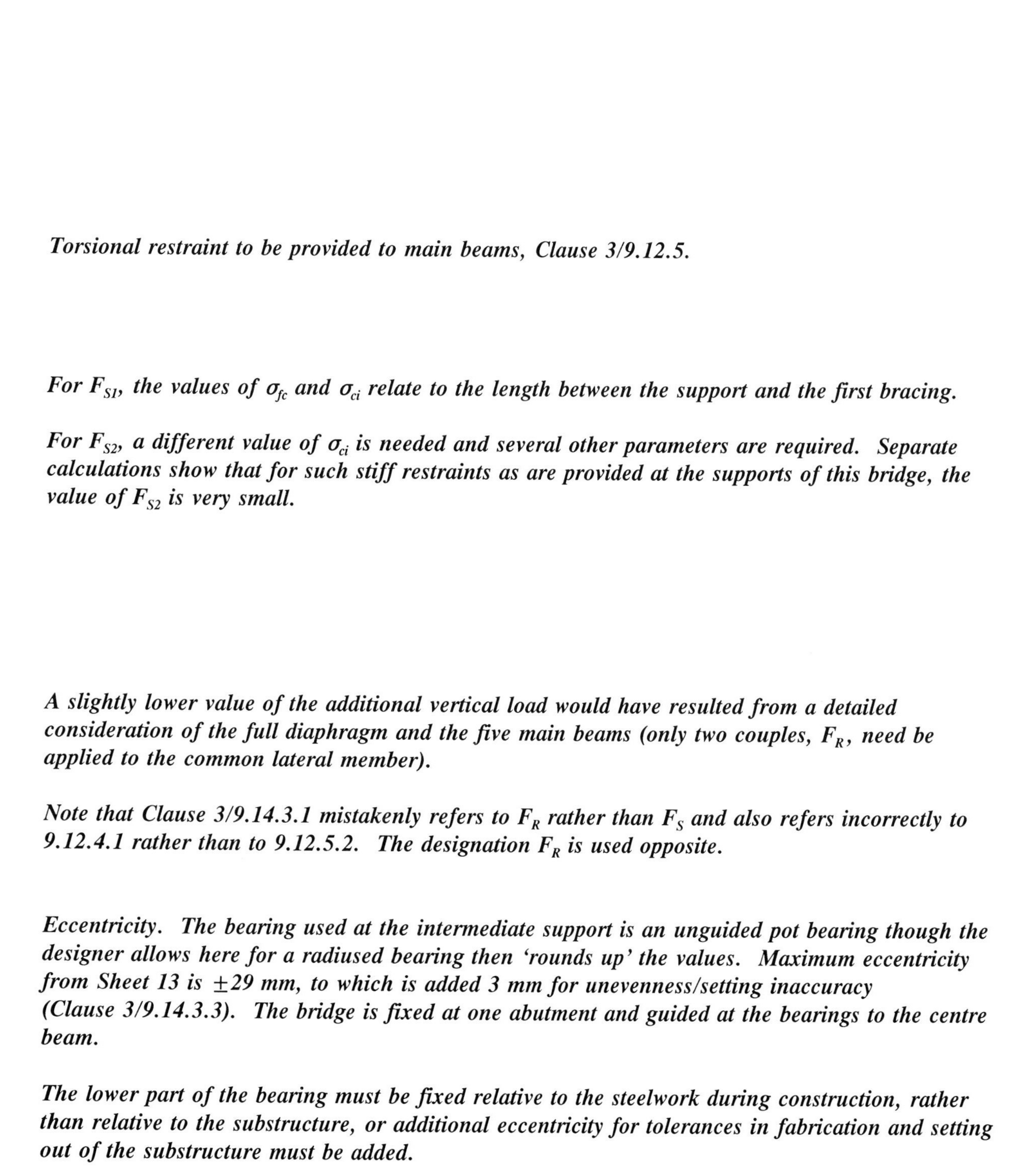

Torsional restraint to be provided to main beams, Clause 3/9.12.5.

For F_{S1}, the values of σ_{fc} and σ_{ci} relate to the length between the support and the first bracing.

For F_{S2}, a different value of σ_{ci} is needed and several other parameters are required. Separate calculations show that for such stiff restraints as are provided at the supports of this bridge, the value of F_{S2} is very small.

A slightly lower value of the additional vertical load would have resulted from a detailed consideration of the full diaphragm and the five main beams (only two couples, F_R, need be applied to the common lateral member).

Note that Clause 3/9.14.3.1 mistakenly refers to F_R rather than F_S and also refers incorrectly to 9.12.4.1 rather than to 9.12.5.2. The designation F_R is used opposite.

Eccentricity. The bearing used at the intermediate support is an unguided pot bearing though the designer allows here for a radiused bearing then 'rounds up' the values. Maximum eccentricity from Sheet 13 is ± 29 mm, to which is added 3 mm for unevenness/setting inaccuracy (Clause 3/9.14.3.3). The bridge is fixed at one abutment and guided at the bearings to the centre beam.

The lower part of the bearing must be fixed relative to the steelwork during construction, rather than relative to the substructure, or additional eccentricity for tolerances in fabrication and setting out of the substructure must be added.

__Inner beam pier bearing stiffener__

Forces acting:

(i) Maximum reaction

$= 215\times1.05 + 930\times1.15 + 535\times1.75 + 15\times1.5 + 1930\times1.3 + 9\times1.2 + 55\times1.2$

(steel) (conc.) (SDL) (footway) (AIL) (settle) (shrink)

$=$ __4840 kN__

(ii) Destabilising effect of the web

Web and bearing stiffener are stabilised by in-plane stiffness of diaphragm

$\therefore F_{wi} = 0$

(iii) Lateral forces F_S

$$F_{s1} = \frac{0.006\,M}{d_f[1 - (\sigma_{fc}/\sigma_{ci})^2]}$$

$M = 18100$ kNm $\quad \sigma_{fc} = 254$ N/mm² (Sheet 16) $\quad$ Take $D = 2030$ mm

$$\sigma_{ci} = \frac{\pi^2 ES}{\lambda_{LT}^2} = \frac{\pi^2 \times 205000 \times 1.25}{51^2} = 972 \text{ N/mm}^2$$ (λ_{LT}, S from values on Sheet 18)

$$F_{S1} = \frac{0.006 \times 18100 \times 10^6}{2030\,[1 - (258/972)^2]} = \underline{58 \text{ kN}}$$

Take $F_{S2} = 0$

Bridge is not skew, so $F_{S3} = F_{S4} = 0$

(iv) Vertical forces due to cross-beam action of diaphragm

Extra vertical load

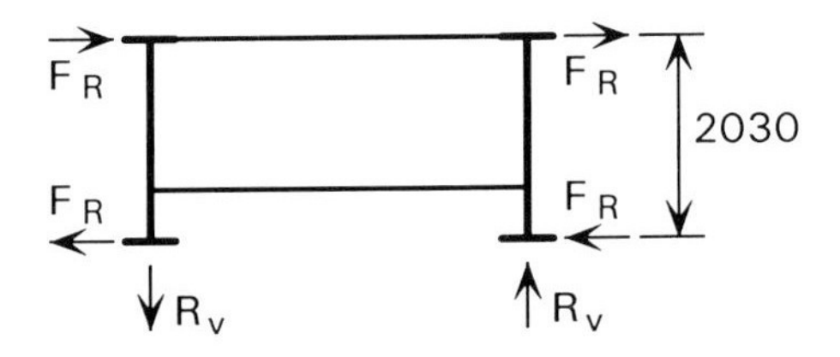

$$= \frac{2 \times F_R \times 2030}{3200} = \frac{2 \times 58 \times 2030}{3200}$$

$=$ __74 kN__

(v) Bending arising from eccentricity of bearing reaction

(a) Movement of beam due to temperature changes:

± 32 *mm longitudinally,* ± 5 *mm transversely*

(b) Changes in contact at spherical surface of bearing due to slope of beam when deflected by load

say ± 5 *mm longitudinally and transversely*

(c)(d) Radiused upper bearing resting on radiused lower part – allow ± 3 *mm*

→ *say eccentricities* *Longitudinal* $= \pm 50$ mm

Transverse $= \pm 20$ mm

Commentary to calculation sheet

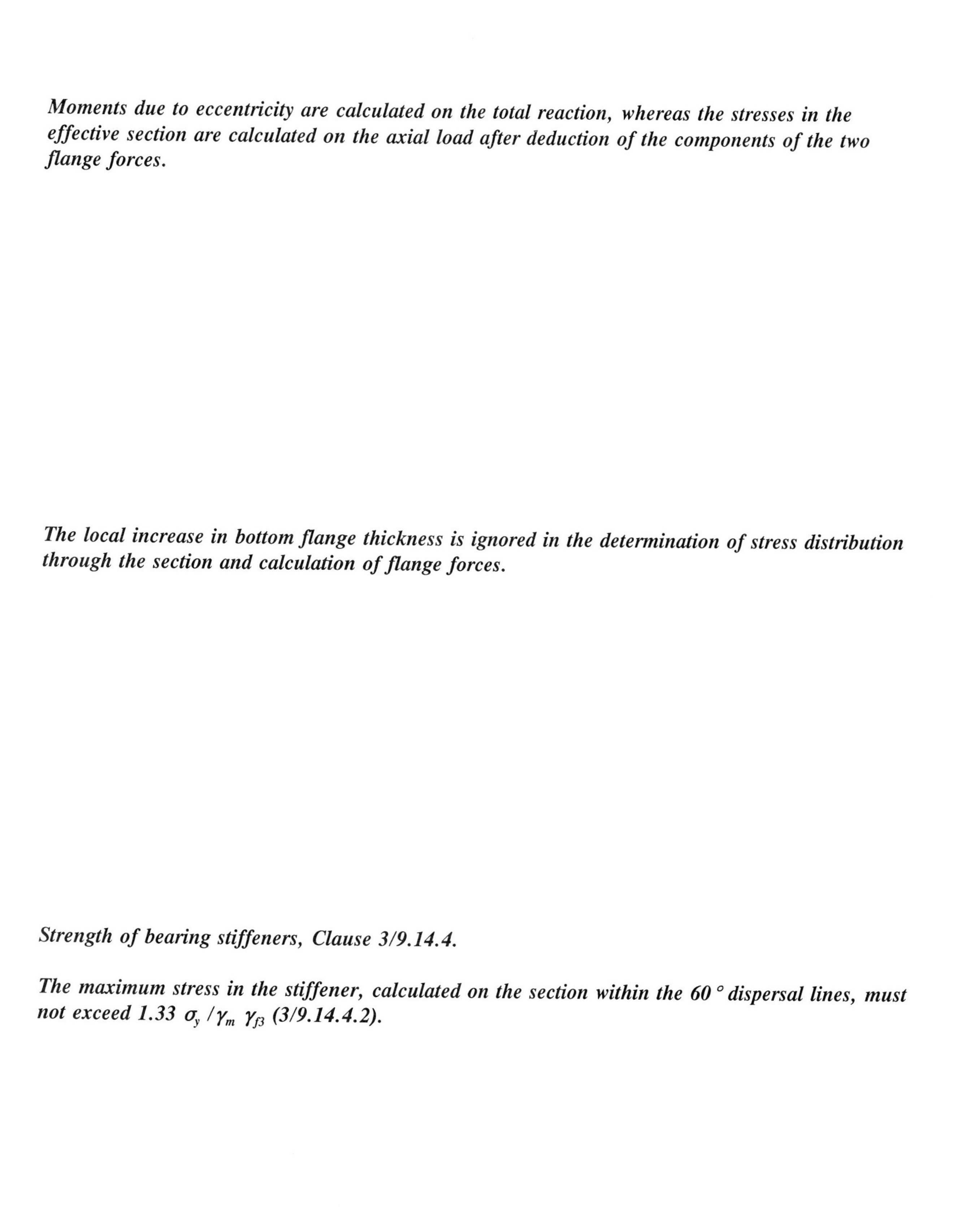

Moments due to eccentricity are calculated on the total reaction, whereas the stresses in the effective section are calculated on the axial load after deduction of the components of the two flange forces.

The local increase in bottom flange thickness is ignored in the determination of stress distribution through the section and calculation of flange forces.

Strength of bearing stiffeners, Clause 3/9.14.4.

The maximum stress in the stiffener, calculated on the section within the 60 ° dispersal lines, must not exceed 1.33 $\sigma_y / \gamma_m \gamma_{f3}$ *(3/9.14.4.2).*

Moment due to longitudinal eccentricity

$= 0.05 \times 4840 = \underline{242 \text{ kNm}}$

Moment due to transverse eccentricity

$= 0.02 \times 4840 = \underline{97 \text{ kNm}}$

(vi) *Vertical component of compression flange forces*

From global analysis, stress in flanges coexistent with max. reaction = 200 N/mm²

∴ *vertical component of two flange forces*

$$= 2 \times 200 \times 600 \times 50 \times \frac{790}{(790^2 + 9125^2)^{1/2}} \times 10^{-3} = \underline{1035 \text{ kN}}$$

Stresses in bearing stiffener:

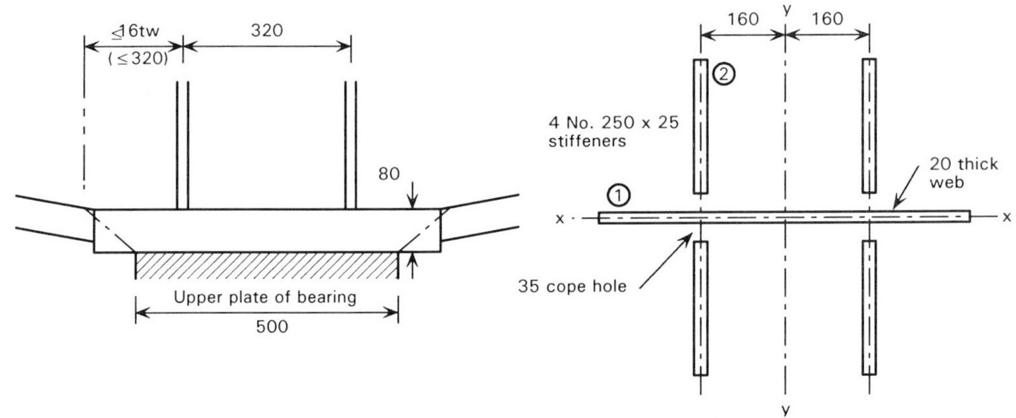

Length of bearing on web $= 500 + 2 \times 80 \tan 60°$

$= \underline{777 \text{ mm}}$

Vertical load on area in bearing

$= 4840 - 1035 = 3805 \text{ kN}$

$A = 777 \times 20 + 4 \times 215 \times 25 = 37040 \text{ mm}^2$

$I_{xx} = 1/12\,(215^3 \times 25 \times 4 + 20^3 \times 777) + 4 \times 215 \times 25 \times 1525^2 = 5.83 \times 10^8 \text{ mm}^4$

$I_{yy} = 1/12\,(777^3 \times 20 + 25^3 \times 215 \times 4) + 4 \times 215 \times 25 \times 160^2) = 1.333 \times 10^9 \text{ mm}^4$

$$\sigma_1 = \frac{3805}{37.04} + \frac{242 \times 389}{1333} = \underline{173 \text{ N/mm}^2}$$

$$\sigma_2 = \frac{3805}{37.04} + \frac{97 \times 260}{583} + \frac{242 \times 160}{1333} = \underline{175 \text{ N/mm}^2}$$

Max. stiffener stress $= 175 \text{ N/mm}^2 < 1.33 \times 299 \text{ N/mm}^2$ *OK*

Commentary to calculation sheet

The maximum stress on the effective section, using the full 16 t_w each side of the stiffener, must not exceed $\sigma_y / \gamma_m \ \gamma_{f3}$ in either web or stiffener.

See Sheet 39 for configuration of diaphragm.

If the beam were restrained by a system of cross-bracing, checks against buckling would be required in accordance with Clause 3/9.14.4.3.

Consider stresses in effective stiffener at base of diaphragm. Axial load is reduced by shear in bottom of web and bending is increased by effects of F_R

$$P = (3805 + 74) \times \left(\frac{1900 - 270}{1900}\right) = \underline{3328\ kN}$$

$$M_y = 242 \times \left(\frac{1900 - 270}{1900}\right) = \underline{208\ kNm}$$

$$M_x = 97 \times \left(\frac{1900 - 270}{1900}\right) + 58 \times 0.31 = \underline{102\ kNm}$$

(lever arm to middle of 80 plt)

At stiffener outstand:

$$\sigma = \frac{3328}{44.2} + \frac{208 \times 160}{2116} + \frac{102 \times 260}{586}$$

$$= \underline{136\ N/mm^2 < 299\ N/mm^2} \quad \underline{OK}$$

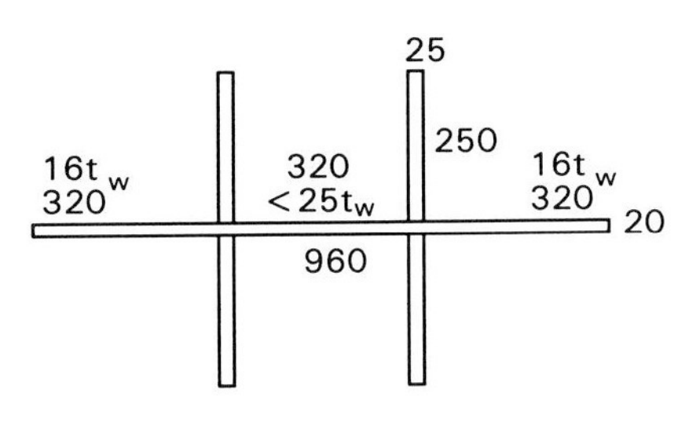

Effective bearing stiffener:

A_{se} = 44200 mm^2
I_{xx} = 5.86 × 10^8 mm^4
r_{se} = 115 mm
I_{yy} = 2.116 × 10^9 mm^4

Buckling of stiffener:

One plane of stiffener plates is restrained by the in-plane connection to the diaphragm

The other plane is restrained at the level of the flanges of the diaphragm, and the worst possible l_e *would then be 1900 - 270 = 1630 mm*

$$\text{and } \frac{l_e}{r} = \frac{1630}{115} = 14$$

Stiffener satisfactory against buckling by inspection

Commentary to calculation sheet

Where the flanges are not parallel, the longitudinal shear flow is given by the expression $\frac{VA\bar{y}}{I} + M\frac{d}{dx}\left(\frac{A\bar{y}}{I}\right)$. *In hogging regions where the depth decreases away from the support, the second term is of opposite sign to the first term (and may conservatively be ignored). Expressing this more simply, the flange forces do not decrease as quickly (away from the support) in a tapered beam.*

Capacity of fillet welds, Clause 3/14.6.3.11.

Value of γ_m *(= 1.2) is given in 3/Table 2.*

13. LONGITUDINAL SHEAR

Flange/web welds at pier

$$\text{Longitudinal shear} = \left(\frac{A\bar{y}}{I}\right) \times \text{vertical shear}$$

where: A = area of section above shear plane being considered
$\bar{y}$ = eccentricity of centroid of area A from neutral axis of section
I = moment of inertia of section

Steel only: top flange $\frac{A\bar{y}}{I} = \frac{500 \times 40\,(1970 - 888)}{57.67 \times 10^9} = 0.375 \times 10^{-3}\ mm^{-1}$

bottom flange $\frac{A\bar{y}}{I} = \frac{600 \times 50\,(888 - 25)}{57.67 \times 10^9} = 0.449 \times 10^{-3}\ mm^{-1}$

Composite cracked: top flange $\frac{A\bar{y}}{I}$

$$= \frac{10057\,(2192 - 1121) + 10057\,(2093 - 1121) + 20000\,(1970 - 1121)}{83.49 \times 10^9}$$

$= 0.449 \times 10^{-3}\ mm^{-1}$

bottom flange $\frac{A\bar{y}}{I} = \frac{3000\,(1121 - 25)}{83.49 \times 10^9} = 0.394 \times 10^{-3}\ mm^{-1}$

Worst shear is on river side of the pier:

ULS shears – steel only: $V = 1.05 \times 115 + 1.15 \times 500 = 696$ kN
(steel) (concrete)

composite: $V = 1.75 \times 290 + 1.5 \times 10 + 1.3 \times 1235 + 1.2 \times 5$
(SDL) (footway) (AIL) (settlement)
$= 2130$ kN

Longitudinal ULS shears

top flange $0.375 \times 696 + 0.449 \times 2130$ = <u>1220 N/mm</u>

bottom flange $0.449 \times 696 + 0.394 \times 2130$ = <u>1150 N/mm</u>

Use simple method of assessing fillet weld capacity.

$$\tau_D = \frac{k\,(\sigma_y + 455)}{\gamma_m\,\gamma_{f3}\,2\sqrt{3}} = \frac{0.9\,(\text{side}) \times (335 + 455)}{1.2 \times 1.1 \times 2 \times \sqrt{3}}$$

$= 155\ N/mm^2$

Min. throat thickness = 1220×0.5 (two welds) / 155
= 3.94 mm

Min. leg length = 5.57 mm

Say allow maximum root gap of 1 mm → <u>8 mm fillet welds</u> <u>OK</u>

Commentary to calculation sheet

Shear connectors need not be considered at ULS, because the section is not compact, and there is no uplift (Clause 5/6.3.4).

Values of $\frac{A\bar{y}}{I}$ *are calculated from properties given on Sheet 9.*

Design strength of shear studs, Clause 5/5.3.3.6.

The design adopted a universal spacing of connectors in the longitudinal direction, the same as the spacing of the transverse reinforcement.

Detailed calculation for $\frac{A\bar{y}}{I}$ *not shown.*

Shear connectors at pier

SLS governs static strength of shear connectors

SLS shear on river side: long term $1.2 \times 290 + 1 \times 5 = 353$ kN

short term $1.1 \times 1235 + 1 \times 10 = 1369$ kN

Considering uncracked sections, and conservatively ignoring shear lag:

$$\text{long-term } \frac{A\bar{y}}{I} = \frac{(242 \times 250)(2140 - 1403) + (45 \times 25)(2002 - 1403)}{114.7 \times 10^9}$$

$$= 0.395 \times 10^{-3}$$

$$\text{short-term } \frac{A\bar{y}}{I} = \frac{(485 \times 250)(2140 - 1618) + (91 \times 25)(2002 - 1618)}{138.6 \times 10^9}$$

$$= 0.463 \times 10^{-3}$$

Shear flow $= 353 \times 0.395 + 1369 \times 0.463 = 634$ N/mm

Use 19 mm ϕ studs. Min. height $= 25 + 45 + 25 + 20 + 40 + 15 = 170$ mm

(haunch + cover + T25 + T20 + clear + head)

Use 175 mm long studs

Nominal static strength = 109 kN (Grade 40 concrete, height over 100 mm)

Design strength = 109 / 1.85 = 59 kN

For studs @ 150 mm crs – shear per group = $634 \times 150 \times 10^{-3} = 95.1$ kN

∴ use 2 No. studs @ 150 mm crs (118 kN > 95 kN OK)

Shear connectors at shallow end of haunch (Node 11)

SLS shear: long term $1.2 \times 160 + 1.0 \times 5 = 197$ kN

short term $1.1 \times 715 + 1.0 \times 10 = 797$ kN

Long-term $\frac{A\bar{y}}{I} = 0.693 \times 10^{-3}$ mm^{-1} Short-term $\frac{A\bar{y}}{I} = 0.756 \times 10^{-3}$ mm^{-1}

Shear flow $= 197 \times 0.693 + 797 \times 0.756 = 739$ N/mm

Shear per 150 mm $= 150 \times 739 \times 10^{-3} = 111$ kN

∴ use 2 No. studs @ 150 mm crs

Shear connectors at abutment

SLS shear: long term $1.2 \times 75 + 1.0 \times 4 = 94$ kN

short term $1.1 \times 1007 + 1.0 \times 3 = 1110$ kN

Long-term $\frac{A\bar{y}}{I} = 0.699 \times 10^{-3}$ mm^{-1} Short-term $\frac{A\bar{y}}{I} = 0.765 \times 10^{-3}$ mm^{-1}

Shear flow $= 94 \times 0.699 + 1110 \times 0.765 = 915$ N/mm

Shear per 150 mm $= 150 \times 915 \times 10^{-3} = 137$ kN

Commentary to calculation sheet

Transfer of shear flow at the ends of beams, Clause 5/5.4.2.3.

Some designers also calculate the extra shear flow due to a change of primary shrinkage stresses at a change of beam section, but the values are usually small and are neglected.

Calculations for shear flow show that pairs of studs at 150 mm centres are adequate over the full length, except immediately adjacent to the abutments. The groups of three studs required there must be provided over at least 10% of the span (= 2.4 m). Separate calculations show that the shear 2.4 m from the abutment is low enough to change to groups of two.

Transverse reinforcement is designed at ULS, in accordance with Clause 5/6.3.3.

The designer has simply checked that the reinforcement is adequate to transfer the capacity of the shear studs, rather than determine the actual ULS shear flows.

Longitudinal shear due to shrinkage:

Stresses in concrete $-0.4\ N/mm^2$ *at top of slab*

$-0.8\ N/mm^2$ *at bottom of haunch (from Sheet 14)*

Force $Q = 250 \times 3200 \times 0.60 + 500 \times 25 \times 0.8 = \underline{490\ kN}$

Length over which Q to be transferred, $l_s = 2\sqrt{\dfrac{KQ}{\Delta f}} = 2\sqrt{\dfrac{0.003 \times 490 \times 10^3}{200 \times 10^{-6}}} = 5420\ mm$

Shear flow $= \dfrac{490 \times 10^3}{5420} \times 2 = 181\ N/mm$

This is in the opposite direction to the DL + LL shear, so does not govern

Longitudinal shear due to positive differential temperature (combination 3 loading)

Force in slab $Q = 94\ kN$ *(from calculation for differential temperature – see Sheet 14)*

$$l_s = 2\sqrt{\frac{KQ}{\Delta f}}$$

$k = 0.003\ mm^2\ N$ *(normal density concrete)*

$\Delta f = 5.4 \times 1.2 \times 10^{-5} = 6.48 \times 10^{-5}$

$$l_s = 2\sqrt{\frac{0.003 \times 94 \times 10^3}{6.48 \times 10^{-5}}} = 4170\ mm$$

Shear flow $= \dfrac{94 \times 10^{-3}}{4170} \times 2 = 45\ N/mm$ *(max.)*

This is much smaller than shear flow due to HB, so combination 1 governs

Use *3 No. studs @ 150 mm crs* *(resistance 180 kN > 137 kN OK)*

Transverse reinforcement

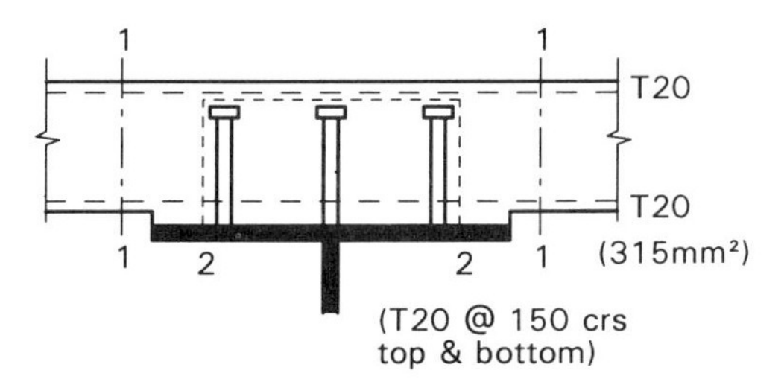

Length of shear plane 1 $= 500\ mm$ *(2 No.)*

Length of shear plane 2 $= 175 + 330 + 175 = 680\ mm$

Resistance on plane 1:

lesser of $0.9 \times 500 + 0.7 \times 460 \left(\dfrac{4 \times 314}{150}\right) = 3150\ N/mm$

and $0.15 \times 500 \times 40 = 3000\ N/mm$

Resistance on plane 2:

lesser of $0.9 \times 680 \times 0.7 \times 460 \left(\dfrac{2 \times 314}{150}\right) = 1960\ N/mm$

and $0.15 \times 680 \times 40 = 4080\ N/mm$

Resistance of 3 No. studs at ULS

$= \dfrac{1}{1.4} \times 109 \times 3 = 234\ kN$

Per mm: 1560 N/mm

Resistance $= \underline{1960\ N/mm > 1560\ N/mm}$ *OK*

Commentary to calculation sheet

The bridge was designed with site splices in the main beams, but for the purposes of this example the more extensive calculations needed for a bolted splice are given.

Calculations are given for worst sagging moment at node 11. Other calculations for worst hogging, with cracked slab, generally give lower steel stresses.

Capacity of HSFG bolts, Clause 3/14.5.4. Values of γ_m from Table 2(b), Part 3.

If the splice were not located close to a bracing position, the design load for the compression flange would have to be increased in accordance with Clause 3/14.4.2.1.

Bolt spacing, Clause 3/14.5.1.1.

14. SPLICE DESIGN

Consider main span splice (shallow end of haunch):

Shear: ULS (Sheet 21) = 1590 kN

SLS $45 + 270 + 1.2 \times 160 + 10 + 1.1 \times 715 + 5$ = 1310 kN

(steel) (conc) (SDL) (foot) (HB) (settle)

Moments (Sheet 15)

ULS steel only -95×1.15 *(concrete)* = –110kNm

composite long $50 \times 1.15 + 1.75 \times 305 + 1.2 \times 53$ = 655 kNm

(conc) (surface) (settle)

composite short 1.3×2565 (AIL) $+ 1.5 \times 25$ *(footway)* = 3370kNm

Hence bending stresses:

$$\sigma_b = \left(\frac{110}{31.3}\right) - \left(\frac{655}{41.2}\right) - \left(\frac{3370}{42.9}\right) = -91\ \text{N/mm}^2$$

$$\sigma_t = -\left(\frac{110}{21.5}\right) + \left(\frac{655}{138.1}\right) + \left(\frac{3370}{343.3}\right) = 9\ \text{N/mm}^2$$

SLS steel only = –95 kNm

composite long $50 + 1.2 \times 305 + 53$ = 469 kNm

composite short $1.1 \times 2565 + 25$ = 2850 kNm

hence bending stresses

$$\sigma_b = \left(\frac{95}{31.3}\right) - \left(\frac{469}{41.2}\right) - \left(\frac{2850}{42.9}\right) = -75\ \text{N/mm}^2$$

$$\sigma_t = -\left(\frac{95}{21.5}\right) + \left(\frac{469}{138.1}\right) + \left(\frac{2850}{343.3}\right) = 7\ \text{N/mm}^2$$

Assume M24 bolts. Friction capacity given by:

$$P_D = k_h \frac{F_v \cdot \mu \cdot N}{\gamma_m \cdot \gamma_{f3}}$$

where: k_h = 1.0 as normal size holes used

F_v = 207 kN proof load BS 4593, no tension

μ = 0.50 for surfaces sprayed with aluminium

γ_m = 1.2 (SLS) = 1.3 (ULS)

N = 2 as cover plates on both faces

γ_f = 1.0

Hence P_D = 173 kN at SLS and 159 kN at ULS

Design to friction at SLS

Number of bolts required at bottom flange $= \dfrac{600 \times 45 \times 75 \times 10^{-3}}{173}$ = 12 bolts

Use 3 rows of 6 bolts

Number of bolts required at top flange $= \dfrac{500 \times 30 \times 7 \times 10^{-3}}{173}$ = 1 bolt

Use 2 rows of 4 bolts

Commentary to calculation sheet

If the resultant force on the top bolt were slightly too great, the web splice could be designed to carry only the moment in the web over the depth of the covers. The load at the top and bottom of the web would then have to be carried by the flange splices.

A_{eq} *conservatively taken as tensile area.*

General requirements for cover plates are given in Clause 3/14.4.1. Here the designer chose cover plates of gross sectional area approximately equal to each element (flange or web) being spliced. Note that in the bearing check on outer plies, a lower k_3 *factor applies. Here the bearing is adequate by inspection.*

Maximum bolt hole sizes, Clauses 3/14.5.3.1 and 6/4.5.

Design stresses in cover plates in tension, Clause 3/14.4.3.

The Steel Construction Institute

Silwood Park, Ascot, Berks SL5 7QN
Telephone: (01344) 623345
Fax: (01344) 622944

CALCULATION SHEET

Job No: *BCR825*	Sheet *36* of *43*	Rev *B*
Job Title *Design of composite bridges – Worked example no. 2*		
Subject *Splice design*		
Client *SCI*	Made by *DCI*	Date *Sep 2000*
	Checked by *CRH*	Date *Dec 2000*

SLS forces in web:

Shear = *1310 kN*

Axial load = $\frac{1}{2}(75 - 7) \times 1125 \times 16 \times 10^{-3} = 612$ *kN*

Moment = $41 (1/6 \times 1125^2 \times 16 \times 10^{-6}) + 0.05 \times 1310 = 204$ *kNm*

(0.05 lever arm between centre of bolts group and centreline web splice)

Min. spacing of bolts = 25 × bolt diameter = 60 mm. Choose 15 No. bolts @ 65 crs

Bolt group Z $= 2(65^2 + 130^2 + 195^2 + 260^2 + 325^2 + 390^2 + 455^2) \times \dfrac{1}{455}$

$= 2600$ *mm*

Forces on extreme web bolt due to:

Shear = *1310 / 15* = *87.3 kN*

Axial = *612 / 15* = *40.8 kN*

Moment = *204 / 2.60* = *78.5 kN*

Resultant = $[87.3^2 + (40.8 + 78.5)^2]^{1/2}$ = ***148 kN < 173*** ***OK***

Check at ULS

Shear capacity of bolts = $n\, A_{eq}\, \sigma_q / \gamma_m\, \gamma_{f3} \sqrt{2}$ *3/14.5.3.5*

(double shear) = $\dfrac{2 \times 353 \times 635}{1.1 \times 1.1 \times \sqrt{2}} \times 10^{-3}$ = ***262 kN***

Bearing capacity inner ply = $k_1\, k_2\, k_3\, k_4\, \sigma_y\, A_{cb} / \gamma_m\, \gamma_{f3}$ *3/14.5.3.6*

Check for 16 mm web: Capacity = $1.0 \times 1.7 \times 1.2 \times 1.5 \times 355 \times 24 \times 16 / 1.05 \times 1.1$

= ***361 kN***

Load in bottom flange

$= 600 \times 40 \times 91 = 2180$ *kN* *Per bolt (18 bolts): 2180 / 18 = 121 kN*

Load in web

Shear = *1590 kN*

Axial $= 41 \times 1125 \times 16 \times 10^{-6} = 738$ *kN*

Moment $= 50 \times 1/6 \times 1125^2 \times 16 \times 10^{-6} + 0.05 \times 1590 = 248$ *kN*

Forces in bolt:

Shear 106.0 kN, axial 49.2 kN, moment 95.4 kN

Resultant $= [106^2 + (49.2 + 95.4)^2]^{1/2} =$ ***179 kN***

OK in bearing/shear

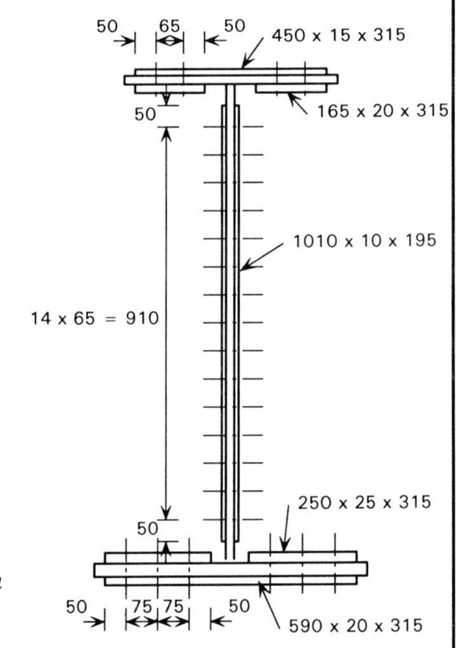

Cover plates

Choose 10 mm for webs (bearing stress less than in web – OK) and 15, 20 and 25 mm on flanges, to roughly match and balance flange areas

Bolt spacings

Min. spacing $= 2.5\, d = 2.5 \times 24 = 60$ *Say 65 mm*

Edge/end distance $= 1.5 d_{hole} = 1.5 \times 26 = 39$ *Say 50 mm*

Max. across web splice $= 2 \times 50 = 100$ *mm*

i.e. < 12 t (120 mm) ***OK***

Commentary to calculation sheet

Restraint force F_R occurring with discrete intermediate restraints, Clause 3/9.12.2.

In service, wind on the live load, the parapet, the edge of the slab and the top half of the girder is carried directly by the deck slab. Only the wind on the bottom half of the girder is carried via the bracing. During erection, the bracing acts to share the lateral load among the five beams.

σ_{fc} *is the maximum stress in the length of flange being restrained.*

If the displacements and rotations at the nodes of the grillage model (which does not include any stiffness for the bracing system, see commentary facing Sheet 7) are imposed on a model of the bracing, conservative values will generally be obtained for the design loads of the bracing system.

The Steel Construction Institute

Silwood Park, Ascot, Berks SL5 7QN
Telephone: (01344) 623345
Fax: (01344) 622944

15. BRACING

Bracing provided at shallow end of haunches

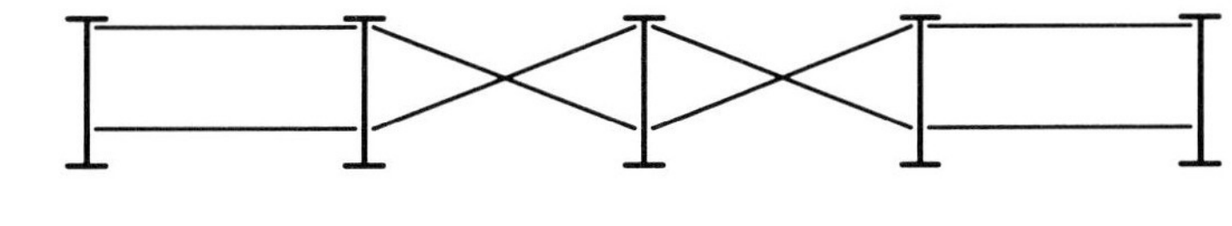

Configuration of bracing members

Loads to be considered:

Lateral restraint to beams
Wind loads
Live loads (global action)

Lateral restraint to beams (combination 1 loading)

$$F = \left(\frac{\sigma_{fc}}{\sigma_{ci} - \sigma_{fc}}\right)\frac{EI_c}{16.7\ell_R^2}$$

σ_{fc} = 254 N/mm² $\lambda_{LT} = 51$ $S = 1.25$ (Sheet 29) σ_{ci} = 972 N/mm² (Sheet 29)

$$F_R = \frac{254 \times 205000 \times 9 \times 10^8}{(972 - 254) \times 16.7 \times 9250^2} = 46 \text{ kN}$$

Or, 1.25% of flange force = $0.0125 \times 600 \times 50 \times 254$ = 95 kN

Use F_R = 95 kN

Wind load (combination 2)
Assume all wind load on windward face
Assume load on bracing from 12 m length, load 1.35 kN/m²

Load due to wind = $1.55 \times 12 \times 1.2 \times ½ = 11$ kN
This is much less than F_R *– ignore combination 2 loading condition*

Live loads
Deflections from the global analysis at the bracing positions are to be applied to a model of the bracing system to determine design forces for the bracing members and their connections

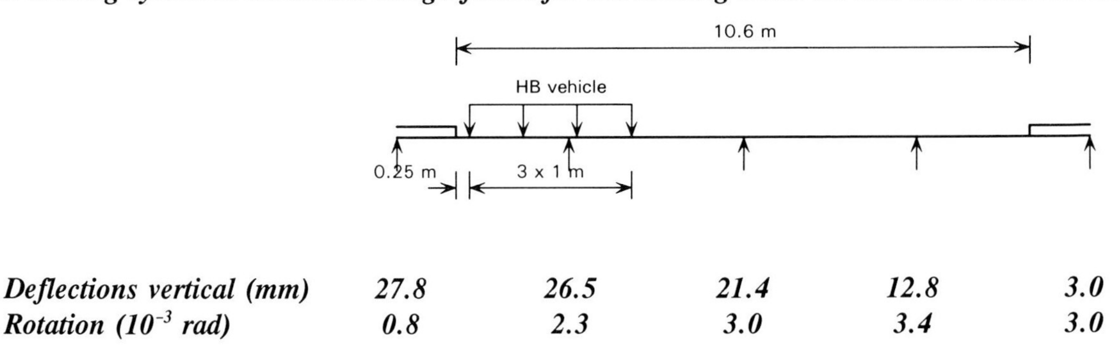

Deflections vertical (mm)	27.8	26.5	21.4	12.8	3.0
Rotation (10^{-3} rad)	0.8	2.3	3.0	3.4	3.0

Commentary to calculation sheet

A simple plane frame model is adequate to determine forces resulting from the displacements, which are applied at deck slab level. The slab could have been modelled, using an appropriate sectional area and inertia.

Axial flexibility of one horizontal = L/AE = 3200 / (3480 × 205000) = 4.5 × 10^{-6} mm. By inspection, the bracing system is stiff enough to act as a fully effective restraint (see comment facing Sheet 17).

The web stiffeners are sufficiently wide for a three-bolt lapped connection for the angle bracing member.

The lateral force F is split into two here to determine its contribution to the design of the lower bracing member.

The two sets of forces in the bracing members given opposite are not coexistent because they relate to different loadcases. Nevertheless, if the bracing is adequate to carry the sum of both effects, it will be satisfactory for the actual coexistent effects.

Calculations are not included here for the detailed checks for the various members of the bracing system and their connections. Checks for fatigue must also be made.

Simplified model of bracing

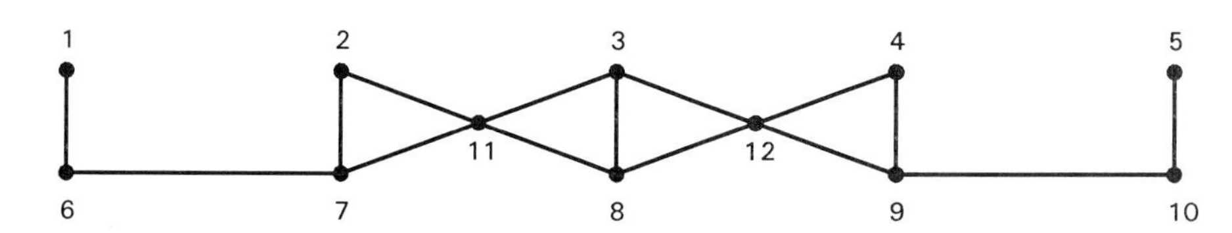

Nodes 1, 2, 3, 4, 5 are conservatively taken as fixed horizontally. Vertical and rotational displacements are imposed on them

Member properties for plane frame analysis

RSA, 150 × 150 × 12 (6–7, 9–10 and diagonals)

$A = 3480\ mm^2$, $I = 7.37 \times 10^6\ mm^4$

Outer girder effective stiffener	A	=	$12650\ mm^2$
	I	=	$9.05 \times 10^7\ mm^4$
Inner girder effective stiffener (similar but with stiffeners on both faces)	A	=	$18900\ mm^2$
	I	=	$2.93 \times 10^8\ mm^4$

<u>Results:</u>

(1) For imposed displacements

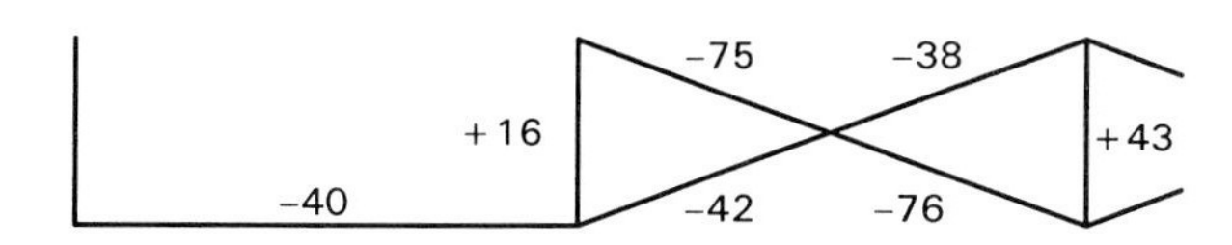

(Forces in kN, +ve compression)

(2) For lateral shears

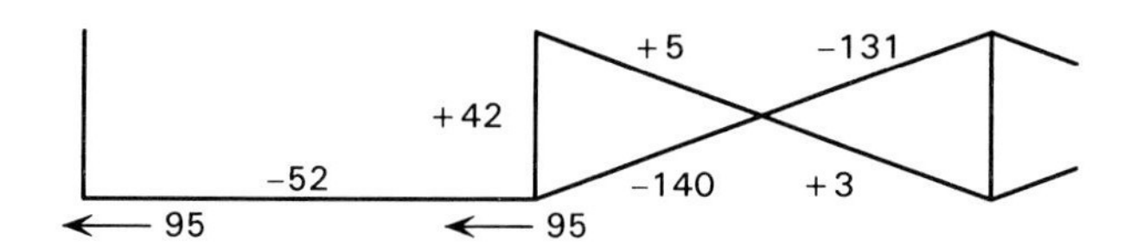

Worst axial load in RSA = –168 kN, σ = –48 N/mm²

Worst BM (member 6–7) M = 7.7 kNm, σ = 114 N/mm²
[coexistent axial load + 18 kN, σ = 5 N/mm²]

Commentary to calculation sheet

The DMRB requires that provision is made for replacement of bearings. The design combines a restraint system for the main beams with transverse beams that can carry the appropriate jacking loads. The splices are site bolted.

The designer has calculated jacking loads to free the outer bearing on the assumption of a simply supported beam between outer and inner beams. It might be necessary to design for the slightly higher loads associated with a continuous transverse beam if it is considered that jacking procedures would not realise this assumption.

16. PIER DIAPHRAGM

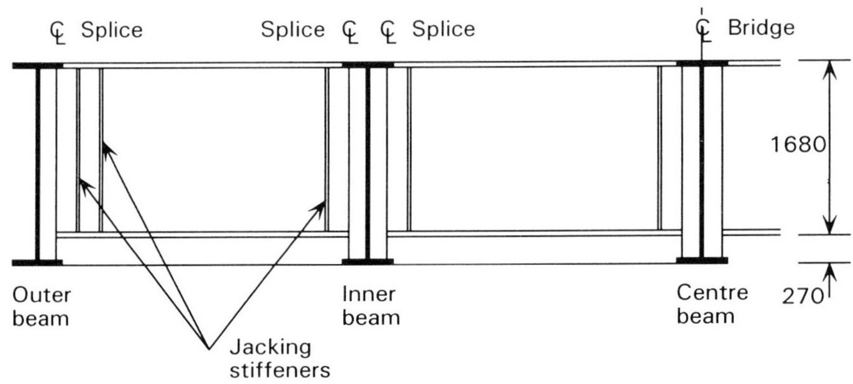

Diaphragm make-up = flanges 400 × 20
web 1640 × 25

Jacking stiffeners provided for replacement of bearings

Section properties for composite section:

In sagging use $2 \times \frac{3200}{8} = 800$ mm *width of slab*

In hogging use $2 \times \frac{3200}{14} = 457$ mm *width of cracked slab (= 3T20 each face)*

(Section properties for these sections are given on Sheet 9)

Two load cases for design of diaphragm:

(i) Sagging under HB vehicle
(ii) Hogging under jacking loads

Sagging: consider 2 HB axles as UDL

$M = 1.3 \times 900 \times 3.2 / 8 = 468$ kNm

Hogging: consider jacking outer beam with HA loading, to relieve reaction at outer bearing

$R = 1.05\times215+1.15\times930+1.75\times208+1.5\times56+1.5\times576+1.2\times9+1.2\times55+1.2\times264$

(steel) (conc) (surface) (foot) (HA) (settle) (shrink) (SDL)

$= 3001$ kN

$R_J = \frac{3001 \times 3.2}{(3.2 - 0.55) + (3.2 - 0.9)} = 1940$

$BM = 3001 \times 0.9 - 1940 \times 0.35 = 2022$ kNm

3001
R_J R_J
550 35

Allow +15% on BM for non-equal jacking loads

$BM = 2330$ kNm

Jacking case governs, by inspection

Bending due to restraining couple $F_S = 58 \times 2.03$

$= 118$ kNm *(F_S from Sheet 29)*

Commentary to calculation sheet

Strictly a check of LTB is needed, but considering the bottom flange of the cross-beam as a compression strut it can be seen that the slenderness is low, and because M_{ult} is more than adequate, the section should be OK.

The jacking stiffener is designed as a bearing support stiffener (Clause 3/9.14). It is common practice to weld a bearing plate to the underside of a cross member that provides a support point for jacking, though it is shown not to be necessary here, provided that the load is carried over an area that is at least 180 mm diameter.

Bearing stresses are checked on the width of section defined by the dispersal lines (see Clause 3/9.14.4.2 and Figures 26 and 27).

Interaction formula (Clause 3/9.14.4.3). As for the check on the bearing stiffener (Sheet 30), the values of Z_x and Z_y should relate to the same extreme fibre. For a cruciform section, two checks should be made, but one is satisfactory by inspection here.

For the check against yield, stresses in the stiffener and the web (and the equivalent stress σ_e) are calculated on the full effective section (not the section defined by the dispersal lines).

Biaxial stresses (σ_1 and σ_{es2}) are both compressive, so combined stresses are OK by inspection. (See Clause 3/9.14.4.)

Shear stress $= \dfrac{3001 \times 10^3}{1640 \times 25} = 73\ N/mm^2$

Shear resistance: $\dfrac{d_{we}}{t_w}\sqrt{\dfrac{\sigma_{yw}}{355}} = \dfrac{1640}{25}\sqrt{\dfrac{345}{355}} = 64.7$

$\phi = 0.25 \quad \therefore \dfrac{\tau_l}{\tau_y} = 0.98$ (3/Figure 12)

Hence resistance $= \dfrac{345 \times 0.98 \times 1}{\sqrt{3} \times 1.1 \times 1.05} = \underline{169\ N/mm^2 > 75\ N/mm^2}$ ***OK***

Bending resistance $M_{ult} = 34.1 \times 345 = 11760$ *kNm*

l/r of bottom flange = 3200 / 115 = 27 *So bending resistance OK by inspection*

Jacking stiffeners

Load = 1940 × 1.15 = 2230 kN (15% *for non-equal loads*)

Allow for eccentricity 10 mm in either direction

BM = 0.01 × 2230 = 22.3 kNm

Assume 180 ϕ *jack ram bearing on underside of flange*

Bearing length along web = 180 + 2 × 20 tan 60 = 250 mm

= *bearing width across stiffeners*

y
170 x 20
25
x x
y

Bearing area = $250 \times 25 + (250 - 25 - 2 \times 40) \times 20 = 9150\ mm^2$

(Allows for 2 No. 40 mm cope holes)

$I_{xx} = \left(\dfrac{250^3 \times 20}{12} - \dfrac{105^3 \times 20}{12}\right) = 24.1 \times 10^6\ mm^4$

$\sigma_b = \dfrac{2230}{9.35} + \dfrac{22.3 \times 125}{24.1} = 354\ N/mm^2 < \dfrac{1.33 \times 345}{1.1 \times 1.05} = 397\ N/mm^2$ ***OK***

Effective section – consider 400 mm of web (= 16 t) each side of stiffener

$A = 2 \times 170 \times 20 + 800 \times 25 = 26800\ mm^2$

$I_{xx} = \dfrac{365^3 \times 20}{12} = 81.0 \times 10^6\ mm^4 \qquad I_{yy} = \dfrac{800^3 \times 25}{12} = 1.07 \times 10^9\ mm^4$

$\ell_s = 1640 \qquad r_{se} = \left(\dfrac{81 \times 10^6}{26800}\right)^{1/2} = 55\ mm$

$\dfrac{\ell_s}{r_{se}} = \dfrac{1640}{55}\sqrt{\dfrac{345}{355}} = 29 \qquad \therefore \sigma_{\ell s} = 345 \times 0.87 = 300\ N/mm^2$ (*curve D of Fig. 37*)

Buckling on middle third ($M = 2/3 \times 22.3 = 14.9$ kNm)

$\dfrac{P}{A_{se}\,\sigma_{is}} + \dfrac{M_{xs}}{z_x\,\sigma_y} + \dfrac{M_{ys}}{z_y\,\sigma} = \dfrac{2230}{26.8 \times 300} - \dfrac{14.9 \times 182.5}{81 \times 345} = 0.49$

$\underline{0.37 < 0.866}$ $(= 1/1.1 \times 1.05)$ ***OK***

Extreme stress in web: $\sigma_{es2} = \dfrac{2230}{26.8} + 22.3 \times \dfrac{400}{1070} = 92\ N/mm^2$

Commentary to calculation sheet

The concrete load includes an allowance for weight of formwork..

Torsional bracing is provided to stabilise the beams against LTB at the wet concrete stage. The bracing at midspan may be removed after erection or left in position; if the latter, it would have to be designed for forces due to live loads.

The triangulated bracing is stiffer than the response of the two main beams, so θ_R is calculated simply on the basis of the stiffness of the beams.

If torsional bracing were inadequate to limit the LTB slenderness sufficiently, plan bracing would be required.

Figure 8 is only plotted up to a value of 1000 for the defining parameter. The value of l_e/L opposite is determined from the expresssion in Annex G.

17. ERECTION CHECK – MIDSPAN MAIN SPAN

Check stability of main span under weight of wet concrete with top flange in compression
Assume torsional bracing between a pair of main beams at three positions – at the shallow ends of haunches and at midspan – treat these as equally spaced for determining ℓ_e *to Clause 3/9.6.4.1.2*

$I_c = 500^3 \times 30 / 12 = 312.5 \times 10^6\ mm^4$

$I_t = 600^3 \times 45 / 12 = 810 \times 10^6\ mm^4$

$I_y = 1.123 \times 10^6\ mm^4 \qquad i = 312.5 / 1122.5 = 0.278$

$r_y = (1.123 \times 10^9 / 60000)^{1/2} = 137\ mm$

For calculation of ℓ_e, $\lambda_F = 42000 / 137 \times (37.5 / 1200) = 9.58$

Hence $v = 0.72$ *(from 3/Table 9)*

To determine θ*, use value of deflection of three-span line-beam model with equal point loads at the three restraint positions*

Deflection per total unit load $= 135 \times 10^{-6}\ mm/N$

Hence, for beam spacing of 3200 mm and unit torque on both beams

$\theta_R = 2 \times (2 \times 135 \times 10^{-6} / 3200^2) = 5.28 \times 10^{-11}\ rad/Nmm$

Parameter for Figure 8 $= v^4 L^3 / EI_c \theta_R d_f^2 = \dfrac{0.72^4 \times 42000^3}{205000 \times 312.5 \times 10^6 \times 5.28 \times 10^{-11} \times 1162^2} = 4360$

Hence $\ell_e / L = 0.324$ *and thus* $\ell_e = 0.324 \times 42000 = 13600\ mm$

For this value of ℓ_e, $\lambda_F = 13600 / 137 \times 37.5 / 1200 = 3.104$ *and thus* $v = 1.14$

For slenderness calculation, determine proportions of bending moment diagram

At supports, moment $= -785 \times 1.05 + -3875 \times 1.15 = -5280\ kNm$

$M_A = M_B = 5280$

At midspan, moment $= 240 \times 1.05 + 1725 \times 1.15 = 2236\ kNm$

$M_M = -5280 - 2236 = -7516$

$M_A/M_B = 1.0 \qquad M_A/M_M = 5280 / -7516 = -0.702 \qquad \eta = 0.57$ *(3/Figure10b)*

$\lambda_{LT} = 13600 / 137 \times 1.0 \times 0.57 \times 1.14 = 65$

Resistance moments

$M_{ult} = 21.5 \times 345 = 7420\ kNm$ *(limited by top flange)*

$M_{pe} = 27.3 \times 335 = 9150\ kNm$ *(using yield of bottom flange)*

Parameter for Figure 11a

$65 \times \sqrt{(345/355) \times (7420/9150)} = 58$

$L/\ell_e = 1/0.324 = 3.09$ *hence* $\ell_w = 42000/3 = 14000\ mm$ *and* $\ell_e / \ell_w = 0.972$

From Figure 11a with these parameters, $M_R / M_{ult} = 0.71$

$M_R = 0.71 \times 7420 = 5270\ kNm$

$M_D = 5270 / (1.1 \times 1.05) = 4560\ kNm > 2236\ kNm$

Commentary to calculation sheet

Fatigue assessment without damage calculation, Clause 10/8.2.

Part 10 specifies that the fatigue vehicle travels within 300 mm of the centre of any lane. Here the designer has used a proportion of the load effects for the HB vehicle, which was positioned within nominal lanes. This generally results in higher stress ranges and is conservative. (It is arguable that the distribution of vehicle paths in a 5 m lane is different from that in a 3.65 m lane, but the code specifies the same positioning in both.)

The impact allowance for loads adjacent to the abutment (Clause 10/7.2.4) has been applied conservatively to the total shear at the end. The distribution shown in Figure 7 of Part 10 reduces linearly to zero at 5 m from any discontinuity.

Stress in welds attaching shear connectors, Clause 10/6.4.

BA 9/81 (as amended in November 1983) recommends allowing for a reduction in fatigue life for plates over 12 mm thick. The reduction is applied here.

It is assumed here that the concrete is cracked, resulting in a slightly higher stress range.

The shear connectors on the top flange introduce a class F detail, but the top flange modulus is larger than the bottom flange modulus, so the top flange is satisfactory by inspection.

18. FATIGUE CHECK

Use simplified method of Part 10, Clause 8.2

Weight of fatigue vehicle = 320 kN – use results of analysis for HB vehicle alone and multiply by $\frac{320}{1800} = 0.178$

Road is 10 m wide, two lane, all purpose. Use Figure 8(c)

<u>Shear connectors at abutment</u>

Max. shear (positive reaction at abutment) due to HB vehicle = 751 kN

Shear due to fatigue vehicle $= 751 \times 0.178 \times 1.25 = 167$ kN (impact)

Min. shear (negative reaction) due to HB vehicle = 100 kN

Min. shear due to fatigue vehicle $= 100 \times 0.178 = 18$ kN

Range of longitudinal shear $= (167 + 18) \times 0.766 = 142$ kN/m *($A\bar{y} / I$ from Sheet 33)*

Load per connector $= 142 \times 0.15 \times ⅓ = 7$ kN

$$\text{Stress range on shear studs} = \frac{\text{load on stud}}{\text{nominal strength}} \times 425$$

$$= \frac{7}{109} \times 425 = 27 \text{ N/mm}^2$$

Limiting stress range from 10/Figure 8(c) with L = 24

$\sigma_H = 47$ N/mm^2

Correction for plate thickness:

For 30 mm plate, factor on design life $= 1.00 - 0.02(30 - 12) = 0.64$

Factor on stress range $= 0.64^{0.125} = 0.95$

Limiting stress range $= 47 \times 0.95 = 44$ N/mm^2 *<u>OK</u>*

<u>Shear connectors at end of midspan haunch (Node 11)</u>

HB vehicle shear range +539 to –90 kN = 629 kN

Range of shear per stud $= (629 \times 0.178) \times 0.757 \times 0.15 \times ½ = 6.4$ kN

$$\text{Stress range per stud} = \frac{6.4}{109} \times 425 = 25 \text{ N/mm}^2$$

Limiting stress range for L = 33 m, $\sigma_H = 43$ N/mm^2

Correction for plate thickness:

For 45 mm plate, factor on design life $= 0.44 - 0.004(45 - 40) = 0.42$

Factor on stress range $= 0.42^{0.125} = 0.90$

Limiting stress range $= 43 \times 0.90 = 39$ N/mm^2 *<u>OK</u>*

<u>Web/bottom flange at sidespan end of haunch (Node 5)</u>

Max. sagging due to HB M = 2138 kNm, with $V \approx 0$

Max. hogging = 1405 kNm, with V = 125

Stresses due to fatigue vehicle, sagging

$$\sigma = 0.178 \times 2138 \times \frac{1}{33.3} = -11.4 \text{ N/mm}^2 \qquad \tau = 0$$

Commentary to calculation sheet

The bracing details, especially the welds to the web, must be checked for fatigue, though the calculations are not included here.

Hogging $\sigma = 0.178 \times 1405 \times \dfrac{1}{33.3} = 7.5\ \text{N/mm}^2$

$\tau = \dfrac{0.178 \times 125}{(1130 \times 16)} = 1.2\ \text{N/mm}^2$

$\sigma_{p\,max} = -11.4$ $\quad \sigma_{p\,min} = 7.5 / 2 + [(7.5 / 2)^2 + 1.2^2]^{0.5} = 7.7\ \text{N/mm}^2$

$\sigma_v = 19.1\ \text{N/mm}^2$

Welding of stiffener to bottom flange and web is *<u>Class F</u>* *(10/Figure 17b)*

Hence from 10/Figure 8(c) with $L = 24$, $\sigma_H = 25\ \text{N/mm}^2$

<u>Stress range of 19.1 N/mm² in web is OK</u>

Correction for plate thickness:

For 40 plate, factor on design life = 0.44

Factor on stress range = $0.44^{1/3}$ = 0.76

Limiting stress range = 25×0.76 = $19\ \text{N/mm}^2$

<u>Stress range of 18.9 N/mm² in flange is OK</u>

<u>Web/bottom flange at pier</u>

HB loads = $M = -2870$ kN with $V = 900$ and $+400$ kN with $V = 0$

Hence fatigue vehicle stresses:

Hogging $\sigma = -0.178 \times 2870 / 73.2 = 7.0\ \text{N/mm}^2$

$\tau = 0.178 \times 900 / (1.9 \times 20) = 4.2\ \text{N/mm}^2$

Sagging $\sigma = 0.178 \times 400 / 73.2 = 1.0\ \text{N/mm}^2$

$\sigma_{p\,max} = (7.0 / 2)^2 + [(7.0 / 2)^2 + 4.2^2] = 9.0\ \text{N/mm}^2$ $\quad \sigma_{p\,min} = -1.0$

$\sigma_v = 10.0\ \text{N/mm}^2$

Welding of stiffeners to flange and web is Class F

From 10/Figure 8(c) with $L = 42$, $\sigma_H = 25\ \text{N/mm}^2$

<u>Stress range of 10.0 N/mm² in web is OK</u>

Correction for 50 mm plate = $0.44 - 0.004\,(50 - 40)$ = 0.40 *on design life*

Factor on stress range = $0.40^{1/3}$ = 0.74

Limiting stress range = 25×0.74 = $18.5\ \text{N/mm}^2$

<u>Stress range of 8.0 N/mm² in flange is OK</u>

<u>Bottom flange at mid mainspan</u>

Max HB *Sagging* = 3060 kNm ($V \approx 0$)

Hogging = 420 kNm ($V = 0$)

$\sigma_v = 0.178 \times 3480 / 42.9 = 14\ \text{N/mm}^2$

Welding of stiffener for midspan bracing is Class F.

From Figure 8(c) with $L = 42$, $\sigma_H = 25\ \text{N/mm}^2$

Correction for 45 mm plate = 0.75 on stress range

Limiting stress range = 25×0.75 = $18.8\ \text{N/mm}^2$

<u>Stress range of 14 N/mm² in flange is OK</u>

Worked Example Number 3

Contents

Commentary to Calculation Sheet

Silwood Park, Ascot, Berks SL5 7QN
Telephone: (01344) 623345
Fax: (01344) 622944

CALCULATION SHEET

Job No:	*BCR825*	Sheet *1* of *7*	Rev *B*
Job Title	*Design of composite bridges – Worked example no. 3*		
Subject	*Outline drawings*		
Client *SCI*	Made by *DCI*	Date *Sep 2000*	
	Checked by	Date	

1. OUTLINE DRAWINGS

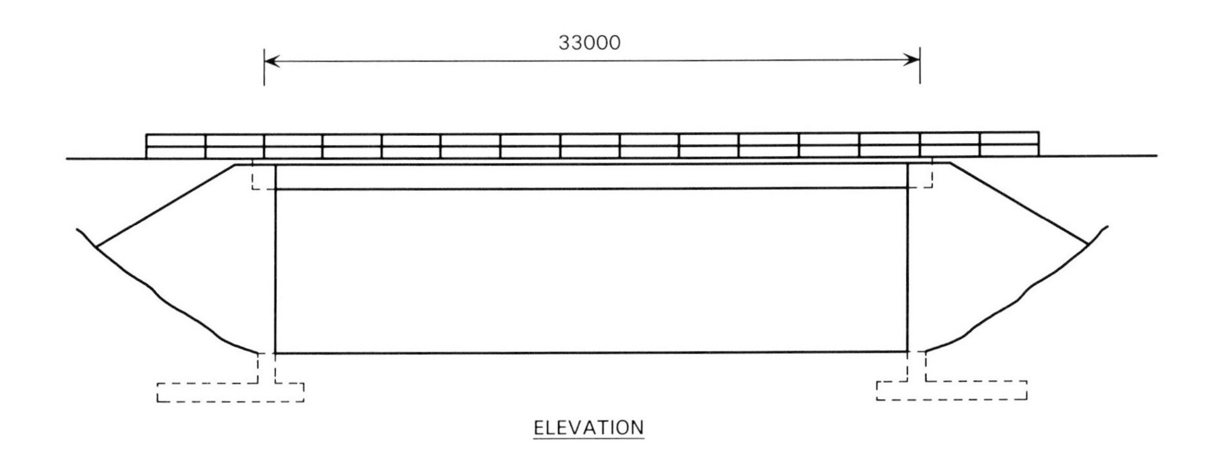

ELEVATION

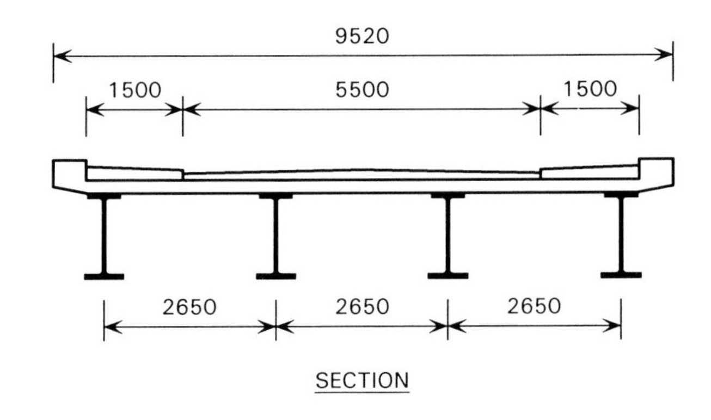

SECTION

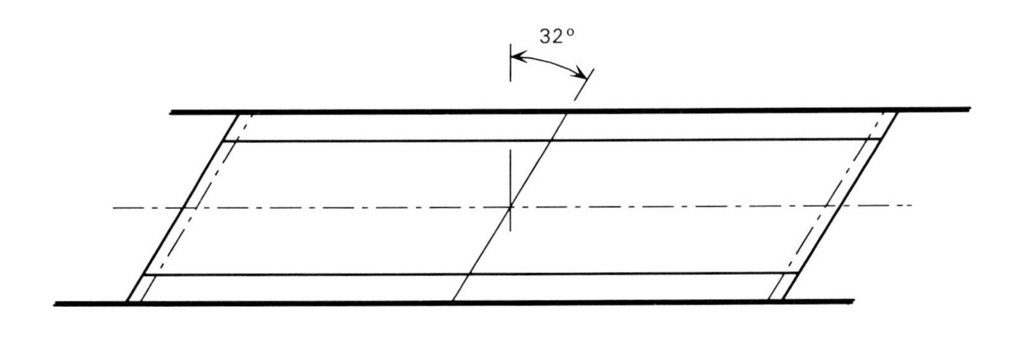

PLAN

Commentary to Calculation Sheet

Transverse bending moments are calculated as the sum of three parts:

(i) Moments due to dead loads

(ii) Distribution moments (from global analysis of live loads)

(iii) Local moments from the application of individual wheel loads between main beams.

The first are generally small and were calculated manually.

The second are taken from computer output.

The third are calculated on Calculation Sheets 3 and 4.

In the midspan region, the sagging distribution moment alleviates the local hogging due to wheels either side of a main beam. Near the ends of the span, the distribution moments will be much less. They may be conservatively taken as zero.

Silwood Park, Ascot, Berks SL5 7QN
Telephone: (01344) 623345
Fax: (01344) 622944

CALCULATION SHEET

Job No: *BCR825*	Sheet *2* of *7*	Rev *B*
Job Title *Design of composite bridges – Worked example no. 3*		
Subject *Deck slab*		
Client *SCI*	Made by *DCI*	Date *Sep 2000*
	Checked by	Date

2. *DECK SLAB*

Transverse bending of slab

(i) *Dead loads*

Nominal loading on a 1 m wide strip of slab across the width of the deck

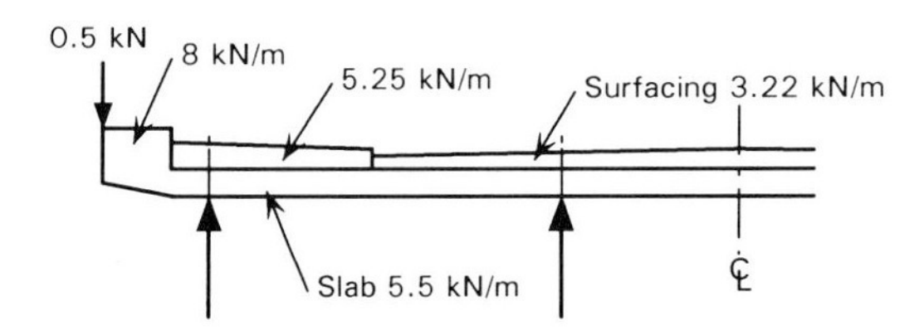

ULS bending moment diagram for above loading (factored by γ_{FL}):

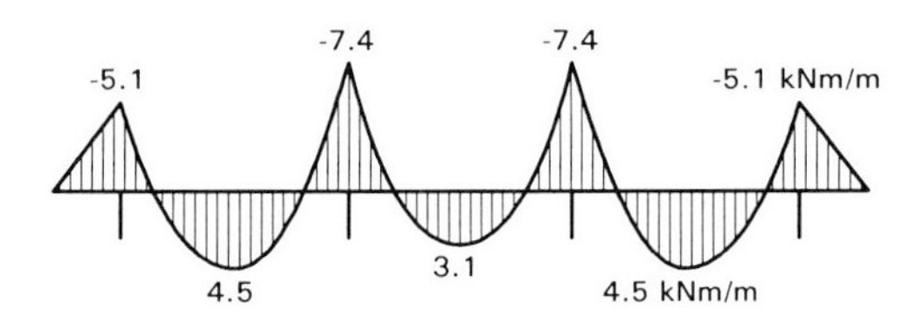

(ii) *Global bending under live loads*

For worst global transverse bending coexistent with the local loading on the slab, sagging between main beams, consider the end of the HB vehicle at midspan

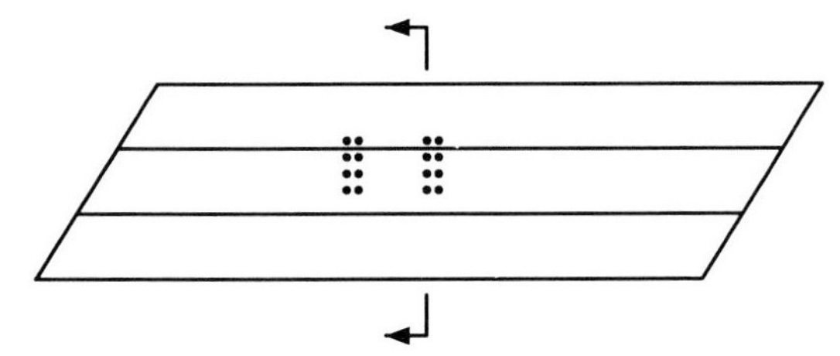

From grillage model, transverse BMD in midspan region is:

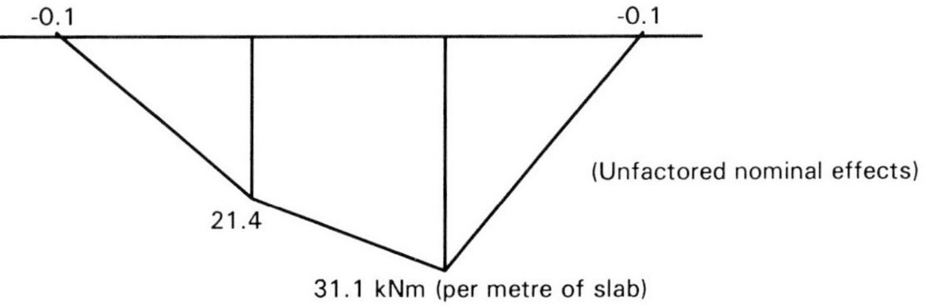

Transverse bending moments at mid span

For worst hogging over main beams, take global transverse bending as zero

Commentary to Calculation Sheet

The third group of moments is calculated manually using Pucher influence charts[1].

A simply supported plate strip is assumed. This is conservative, in view of the continuity of the slab on either side. Restrained edges would underestimate the midspan moment.

Pucher charts assume that Poisson's ratio is zero. Correction for a real value of say 0.2 can be made by using another chart to calculate the orthogonal moment under the same loading and support conditions, multiplying it by 0.2 and adding to the first moment. The correction here would be small and is more than offset by the conservatism of neglecting continuity. If there are only two main beams, the correction should be made.

Contact area of wheel loads is given by Clause 2/6.3.2. The load may be dispersed through the surfacing and the slab (see Clause 2/6.2.6), but the dispersal has been neglected here.

---

[1] *PUCHER, A. Influence surfaces of elastic plates, Springer-Verlag, Vienna, 1964*

Silwood Park, Ascot, Berks SL5 7QN
Telephone: (01344) 623345
Fax: (01344) 622944

CALCULATION SHEET

Job No: *BCR825*	Sheet *3* of *7*	Rev *B*
Job Title *Design of composite bridges – Worked example no. 3*		
Subject *Deck slab*		
Client *SCI*	Made by *DCI*	Date *Sep 2000*
	Checked by	Date

(iii) *Local bending under wheel*

(a) *Sagging midway between beams*

Use Pucher chart for plate-strip with simply supported edges
Consider only the HB wheels between the main beams
Use loaded areas 300 mm diameter

Hence from chart:

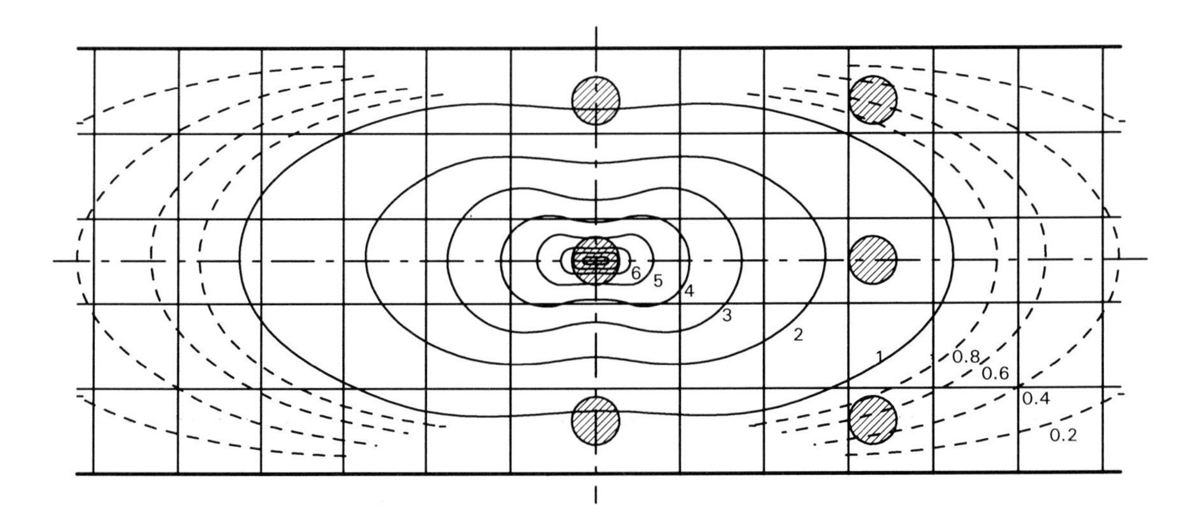

Transverse moment = (load/wheel) $\times \dfrac{1}{8\pi} \times$ *(sum of average ordinates under each wheel)*

Using ordinate values from diagram above:

$$M = \frac{75}{8\pi} + (0.9 + 6.5 + 0.9 + 0.55 + 1.5 + 0.55)$$

$$= 32.5 \text{ kNm/m (nominal effect – unfactored)}$$

Commentary to Calculation Sheet

The four wheels on each axle are placed symmetrically about a main beam. Use is then made of the symmetry by using the chart for a fixed simply-supported plate strip. Continuity beyond the simply-supported edge has little effect on this moment. Correction for Poisson's ratio is small and usually reduces the transverse hogging; it may be neglected.

Local bending under wheel (cont'd)

(b) Hogging over main beams

Use Pucher Chart for plate-strip with one fixed and one simply supported edge
Place HB vehicle symmetrically about main beam
Use loaded areas 300 mm diameter

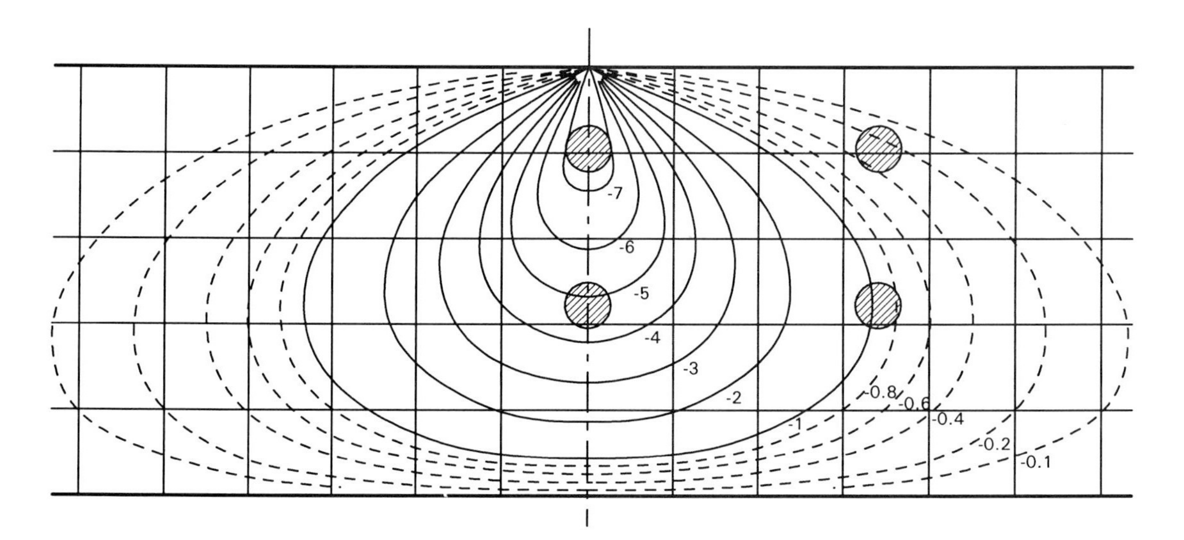

$$\textit{Transverse moment} = \textit{(load/wheel)} \times \frac{1}{8\pi} \times \textit{(sum of average coordinates)}$$

$$= \frac{75}{8\pi} \times (-7.2 - 4.7 - 0.28 - 0.95)$$

$$= 39.2 \text{ kNm/m (nominal effect – unfactored)}$$

Commentary to Calculation Sheet

The highest moments in the slab are transverse, so the transverse reinforcement is placed as the outer layer.

Cover is the nominal cover for Grade 40 concrete given in Part 4 (Clause 4/5.8 and Table 13), plus the additional 10 mm specified by BD 57/95.

The bridge soffit is considered to be in a 'severe' environment and the top surface is in a 'moderate' environment.

Design resistance of slabs, Clause 4/5.3.2.3.

Reinforcement in compression is ignored.

Here γ_{f3} *is put on the strength side, in the same way as for design of the steel parts.*

The Steel Construction Institute

Silwood Park, Ascot, Berks SL5 7QN
Telephone: (01344) 623345
Fax: (01344) 622944

CALCULATION SHEET

Job No: BCR825	Sheet 5 of 7	Rev B
Job Title: Design of composite bridges – Worked example no. 3		
Subject: Deck slab		
Client: SCI	Made by: DCI	Date: Sep 2000
	Checked by:	Date:

Design of transverse reinforcement

ULS moments

Sagging: $3.1 + 1.3 \times 26.3 + 1.3 \times 32.5 = 79.5$ kNm/m

Hogging: $-7.4 + 1.3 \times 0 + 1.3 \times (-39.2) = -58.4$ kNm/m

Try T16s @ 150 crs, top and T20s @ 150 crs, bottom

230 (slab depth); 40 (cover, top); 45 (cover, bottom)

T16 A_s = 1340 mm²/m width

T20 A_s = 2094 mm²/m width

Sagging: $d = 230 - 45 - \frac{20}{2} = 175$ mm

$$M_u = 0.87 f_y A_s Z / \gamma_{f3}$$

$$Z = d\left(1 - \frac{1.1 f_y A_s}{f_{cu} bd}\right)$$

$$= 175 \times \left(1 - \frac{1.1 \times 460 \times 2.094}{40 \times 175}\right) = 149 \text{ mm}$$

$$M_u = 0.87 \times 460 \times 2094 \times 149 \times 10^{-6} / 1.1 = 114 \text{ kNm/m}$$

or $M_u = 0.15 f_{cu} bd^2 / \gamma_{f3} = 0.15 \times 40 \times 1.0 \times 175^2 \times 10^{-3} / 1.1$

$= 167$ kNm/m

Hence M_u = <u>114 kNm/m</u> <u>Satisfactory</u>

Hogging: $d = 230 - 40 - \frac{16}{2} = 182$ mm

$$Z = 182 \times \left(1 - \frac{1.1 \times 460 \times 1340}{40 \times 10^3 \times 182}\right) = 165 \text{ mm}$$

$$M_u = 0.87 \times 460 \times 1340 \times 165 \times 10^{-6} / 1.1$$

= <u>80.4 kNm/m</u> <u>Satisfactory</u>

Commentary to Calculation Sheet

Only 30 units of HB are considered for checking crack widths. Originally, BS 5400 required only 25 units of HB to be considered but this has been increased by BD 24/92 (which implements BS 5400–4) to 30 units. BD16 is due to be brought into line with this when it is updated.

γ_{fl} *is taken as 1.1, in accordance with Part 2 and Clause 5/5.2.6.2 (the original BSI Clause 5/5.2.6.2 permits* γ_{fl} *to be taken as 1.0).*

Calculations of cracked elastic section properties for slab.

Calculation of design crack width, Clause 4/5.8.8.2.

Limiting crack width, Clause 4/4.1.1.1, 'severe' environment.

Silwood Park, Ascot, Berks SL5 7QN
Telephone: (01344) 623345
Fax: (01344) 622944

CALCULATION SHEET

Job No: *BCR825*	Sheet *6* of *7*	Rev *B*
Job Title *Design of composite bridges – Worked example no. 3*		
Subject *Deck slab*		
Client *SCI*	Made by *DCI*	Date *Sep 2000*
	Checked by	Date

Design of transverse reinforcement (cont'd)
SLS crack control

Check sagging midway between beams

DL Say 2.7 kNm/m (approximately; M_{ULS} is 3.1 kNm/m)

LL $1.1 \times (26.3 + 32.5) = 64.7$ kNm/m (Distribution moment of 26.3 kNm from grillage analysis)

DL + LL $= 2.7 + 64.7 = 67.4$ kNm/m

Appropriate $E_c = 31 - \frac{31}{2} \times \frac{2.7}{67.4} = 30.4 \text{ kN/mm}^2$

Modular ratio $\alpha_o = E_s / E_c = 200 / 30.4 = 6.58$

Proportion of reinforcement $r = \frac{A_s}{bd} = \frac{2094}{1000 \times 175} = 0.01197$

Depth to NA $n = d\left(\sqrt{2r\alpha_o + (r\alpha_o)^2} - r\alpha_o\right)$

$= 175 \times [(0.1576 + 0.0062)^{½} - 0.788] = 57.0 \text{ mm}$

$a = d - n/3 = 175 - 57/3 = 156 \text{ mm}$

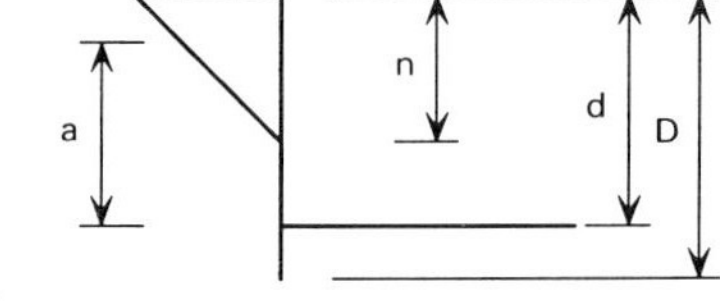

Hence steel stress $= \frac{67.4 \times 10^6}{156 \times 2094} = 206 \text{ N/mm}^2$

Strain $\epsilon = 206 / 200 \times 10^{-3} = 1.03 \times 10^{-3}$

$\epsilon_1 = \epsilon \times \left(\frac{D - n}{d - n}\right) = 1.03 \times 10^{-3} \times \frac{220 - 57.0}{175 - 57.0}$

$= 1.42 \times 10^{-3}$

Since $M_1 > M_1$ $\epsilon_m = \epsilon_1$

Design crack width (BS 5400–4 minimum cover) $= \epsilon_m \times \frac{3\, a_{cr}}{1 + 2\,(a_{cr} - c_{nom})/(h - d_c)}$

45 a 75 75

$a_{cr} = \sqrt{(75^2 + 45^2)} - 20/2 = 77 \text{ mm}$

Commentary to Calculation Sheet

Single HA wheel, Clauses 2/6.2.5 and 2/6.2.6 (γ_{fl} = 1.5).

Critical section for shear under concentrated loads, Clause 4/5.4.4.2.

Punching shear capacity is normally adequate for deck slabs of usual depth and strength (and using v_c appropriate to actual reinforcement, there is a much larger reserve capacity).

Individual HB wheel loads can be checked in the same manner as the HA wheel load, but with γ_{fl} = 1.3 they are not as onerous as the HA wheel load.

Groups of HB wheels should be checked by defining a rectangular critical perimeter around two or more wheels, as prescribed by Clause 4/5.4.4.2. The area would be about 1.9 m × 0.8 m under two wheels or 150 kN; this can be seen to be less severe than around the 100 kN HA load.

Strictly, the maximum shear on a section along the edge of the girder flanges should also be checked (Clause 4/5.4.4.1). Shear forces are required to be calculated from an elastic analysis (4/5.4.1); this should include dead loads, local forces from global distribution and local forces from live loads. The latter can be readily calculated for UDL loads but for a group of local loads (i.e. HB wheels) simple empirical rules must be applied (e.g. spread at 45 ° in plan, share between beams according to statics). Such general shear stresses are normally less severe than local effects; they are not calculated here.

Ultimate shear stress v_c, Part 4, Table 8. (Note that if the actual proportion of reinforcement – over 1% – were used, v_c would be approximately double.)

Value of ξ, Part 4, Table 9. Value for d = 175 mm used here.

Silwood Park, Ascot, Berks SL5 7QN
Telephone: (01344) 623345
Fax: (01344) 622944

CALCULATION SHEET

Job No: *BCR825*	Sheet *7* of *7*	Rev *B*
Job Title *Design of composite bridges – Worked example no. 3*		
Subject *Deck slab*		
Client *SCI*	Made by *DCI*	Date *Sep 2000*
	Checked by	Date

Crack width $= 1.42 \times 10^{-3} \times \dfrac{3 \times 77}{1 + 2(77 - 35)/(220 - 57.0)}$

$= 0.22\ mm < 0.25\ mm$ *Satisfactory*

Concrete stress $= M_p \times \dfrac{2}{a.n.b} = \dfrac{67.4 \times 10^6 \times 2}{156 \times 57.0 \times 1000}$

$= 15.2\ N/mm^2 < 16\ Nmm^2\ (0.4 f_{cu})$ *Satisfactory*

<u>Longitudinal reinforcement</u>

Design of longitudinal reinforcement follows a similar process to that for the transverse reinforcement. Longitudinal moments are generally sagging and of lesser magnitude. Crack width must be checked carefully because of the greater cover to the longitudinal bars.

The abutment trimmer beam provides restraint to the end of the slab; the slab must be checked for hogging close to the trimmer beam.

<u>Shear</u>

Check 'punching' shear under HA single wheel load of 100 kN (Apply $\gamma_{fL} = 1.5$)

Load spread over 300 × 300 mm area at roadway surface
Disperse at 1:2 – loaded area on slab = 400 × 400 mm

Critical perimeter around wheel

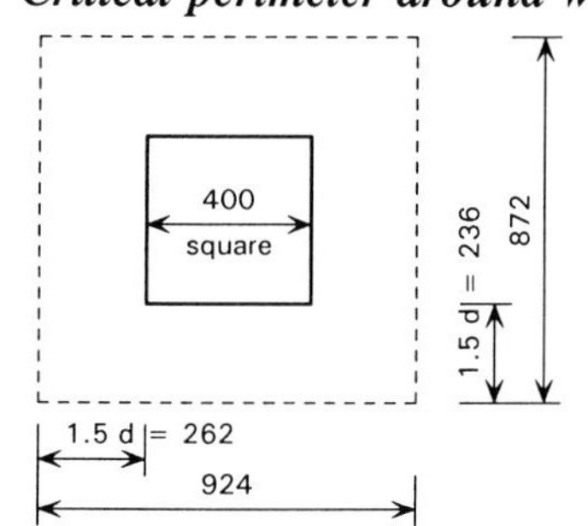

Longitudinal $d = 175$ mm (*Calculation Sheet 5*)

Transverse $d = 230 - 45 - 20 - 16/2 = 157$ mm

Perimeter sectional area $= 2 \times (0.872 \times 0.157 + 0.924 \times 0.175)$

$= 0.597\ m^2$

Shear stress $v = \dfrac{150 \times 10^3}{0.597 \times 10^6} = 0.251\ N/mm^2$

Ultimate shear stress in concrete v_c depends on proportion of reinforcement
But, for Grade 40 concrete, v_c is not less than 0.39 N/mm²

Take ξ_s as 130 ∴ shear capacity $= 1.30 \times 0.39 / 1.1 = 0.46\ N/mm^2$ *<u>Adequate</u>*

Shear around HB wheels, individually and in groups, is satisfactory by inspection